Rain Forest in Your Kitchen

Martin Teitel

Rain Forest in Your Kitchen

THE HIDDEN CONNECTION
BETWEEN EXTINCTION
AND YOUR SUPERMARKET

Foreword by Jeremy Rifkin

ISLAND PRESS

Washington, D.C.   □   Covelo, California

Illustrations on pp. 64–65 by Teka Luttrell, © 1992

Permissions from copyrighted material appear on page 112.

Library of Congress Cataloging-in-Publication Data

Teitel, Martin.
    Rain forest in your kitchen : the hidden connection between extinction and your supermarket / Martin Teitel.
        p.   cm.
    Includes bibliographical references.
    ISBN 1-55963-153-8 (pbk.)
    1. Consumer education.   2. Environmental protection.   3. Biological diversity conservation.   I. Title.
TX335.T388   1992
640'.73—dc20   91-41036
                                                                CIP

Printed on recycled, acid-free paper

Manufactured in the United States of America
10 9 8 7 6 5 4 3 2

CONTENTS

FOREWORD

Eating, more than any other human activity, binds us to the world of nature. It is the vital bridge that connects human culture with the larger environment. The kinds of plants and animals that a society chooses to expropriate and consume provide a mirror to the values and relationships of that society. That is why, in most cultures, eating is celebrated as a sacred act, as well as an act of survival and replenishment.

The late Austrian psychologist Bruno Bettelheim believed that the "eating experience conditions our entire attitude toward the world." The French anthropologist and philosopher Roland Barthes argues that the kinds of agricultural and husbandry practices that a society adopts, the foods it eats, and the methods of preparing those foods serve as a highly sophisticated form of bonding, conveying the entire set of beliefs that make up the distinguishing characteristics of a culture. In this sense, eating is the ultimate political act. Barthes writes, "For what is food? It is not only a collection of products that can be used. . . . It is also, at the same time, a system of communicating; a body of images, a protocol of usages, situations and behaviors."

It has been said that "the history of humankind is only intelligible in the context of the history of food." In traditional agricultural societies, food is viewed as a gift bestowed by the living earth. Nature's fecundity, in turn, depends on a proper relationship being established between culture and the creation. The earth is imbued with sacred value and requires stewardship if it is to willingly give of itself.

In modern society, humanity's relationship to the earth and its bounty has metamorphosed into a strictly utilitarian proposition. The great land masses and oceans, once a shared biological commons, have been systematically enclosed and commercialized, their living inhabitants transformed into private property whose value is measured in purely market terms.

Today, the rich genetic pool of plant and animal life that evolved into complex living communities over millions of years is threatened with extinction at the hands of multinational corporations and market forces. In the name of profits, transnational business firms have rapidly gained commercial control over the earth's seeds and animal breeds, guaranteeing them a virtual monopoly over the world's food supply. They have been aided in their efforts by national governments—especially the United States—that have been quick to extend the patent laws to include ownership of microbes, plants, and animals.

Armed with aggressive marketing strategies and massive advertising budgets, multinational corporations increasingly dictate the terms of the food process from production through consumption. Their policies are leading to a dramatic narrowing of the genetic diversity of the planet and creating an environmental and economic crisis of unparalleled dimensions.

By relying on only a handful of "market-efficient" varieties and breeds for production and marketing around the world, these corporations have greatly diminished the number of plant and animal species under domestication. The human population of the planet is being forced to rely on fewer and fewer varieties and breeds and finds its food choices dangerously narrowed to meet the requirements of mass production, economies of scale, marketing prerequisites, and market performance. If the biological diversity of the planet continues to diminish at the present rate, future generations may find themselves without sufficient genetic reserves to sustain their own survival.

Martin Teitel has written an important book on one of the most critical issues facing our civilization and planet. Restoring biodiversity on the earth and restructuring the agricultural practices and eating habits of Americans are among the most politically important tasks of the coming decades. These are tasks that require not only a change in our personal dietary choices but also a change in our relationship to the natural world from which we draw our sustenance. While Teitel helps us to understand the ecology of genetic diversity and the politics of food, he argues that a change in eating habits must also be accompanied by a personal transformation grounded in a new appreciation of the sacred value of the biological community of which we are a part. Teitel's beautifully written account of our proper relationship to the natural order can help guide all of us in our sojourn toward a more sustainable future. By championing genetic diversity and making a personal choice to change our eating habits to reflect our concern for the larger biological community, we help prepare the ground for a new age in history—one in which we accept our responsibility to be caretakers of the earth.

—Jeremy Rifkin

ACKNOWLEDGMENTS

A large number of people helped produce this small book. I was greatly assisted by readers who offered useful comments on the book's first draft, including Robert Anderson, Anne Fitzgerald, Mary J. Harrington, Michael F. Jacobson, Andrew Kimbrell, Jeremy Rifkin, Sidney Teitel, Shan Thomas, Roxanne Turnage, and Herman E. Warsh.

C. S. Fund staff, interns, and consultants assisted in putting the book together. In particular, Roxanne Turnage gave steady aid and encouragement from the first handwritten draft to the galleys. Dorothy Foster supplied outstanding editorial and organizational support, innovation, and inspiration—draft after draft.

Maryanne Mott and Herman E. Warsh of the C. S. Fund's board provided material, intellectual, and spiritual support every step of the way, as they always do.

The staff at Island Press were uniformly helpful, flexible, and encouraging.

My wife, Mary J. Harrington, was a source of constant support and ideas, as well as illustrating what a real writer can and should be.

Rain Forest in Your Kitchen

INTRODUCTION

The newspaper flung daily onto my front porch reports sometimes on the world biodiversity crisis: "Tropical Rain Forests Disappearing," the headline might say, with the story beneath documenting how some 200,000 acres of rain forest are cleared every day. And, the article might continue, because these forests play host to more than half of all plant and animal species on Earth, their destruction dooms some 4,000 to 6,000 species per year.[1]

But these facts—syndicated and no doubt presented in your daily paper, too—bear so little on our everyday lives that they seem unreal. Tropical rain forests are "over there," in someone else's backyard, beyond our immediate circle of influence. Their preservation is a cause for concern all right, a cause we write checks to support.

Is the biodiversity crisis really so far away? Not at all. It is over our back fences, in the corn, potato, and tomato fields managed by today's corporate American farmers; it is in our backyard gardens, in the commercial, hybrid seeds we plant instead of our family's prized heirloom collection; it is on our refrigerator shelves, in the cosmetically perfect, uniformly sized produce we demand as consumers. Many varieties of food plants and breeds of domesticated animals have recently vanished from America's farms, without eulogy or epitaph. For example:

- More than 6000 known varieties of apples (86 percent of those ever recorded) have become extinct since 1900.
- Since 1900, 2300 pear varieties have disappeared.
- Ninety percent of all chicken eggs sold are laid by one breed, the white Leghorn.
- Seventy percent of the nation's dairy herd is Holstein.
- Two varieties of peas account for 96 percent of the yearly pickings.
- Three varieties of oranges provide 90 percent of Florida's annual harvest.
- Four varieties of potatoes make up nearly 75 percent of the yearly crop, and one variety represents the bulk.

The statistics are astonishing. But the fact that there is a crisis in our backyard is actually good news. Just as our everyday decisions about what to eat can signal disaster for the genetic resources of the planet, they can also forcefully repel further destruction of our ecosystem. Since we all eat to live, we all have multiple opportunities each day to make choices that favor variety in edible plants and domesticated animals. And in today's "global village," what we Americans eat has far-reaching effects. For example, according to a MacArthur Foundation report, "Growing imports of beef by the United States from southern Mexico and Central America during the past twenty-five years has been the major factor in the loss of about half of the tropical forests there—all for the sake of keeping the price of hamburger in the United States about a nickel less than it would have been otherwise."[2] The biodiversity crisis extends across the globe from our kitchens and barbecues.

Perhaps you were already aware that a beefy diet smacks of ecological insensitivity. Most of us are conscious of, if not well read about, the multiple seemingly intractable problems affecting the world today. Often the problems seem so huge, so beyond control, so overwhelming that it is difficult to know what to do.

Most of us do nothing.

This book's purpose is not to stagger you with the specter of biodiversity disaster, but to inspire you with strategies for wielding your consumer power. There are simple, small things that we each can do in the course of everyday living—without altering lifestyle or incurring expenses—that can help to forestall further extinctions and protect biological diversity.

To begin with, we can carry to the grocery store, the fast-food outlet, and the garden center a consciousness of the link between food and the world's genetic base. This link, explained in the first two chapters of this book, is an amalgam of four elements: genetically controlled hybrid and inbred plants, fertilizers and pesticides, corporate agricultural practices, and biotechnology. The strategies for reforging the link are presented in the last six chapters.

The strategies presented in this book are simple; they are meant to give you ideas and inspiration for actions that will help to rescue and conserve our planet's precious genetic resources. None of them require major changes in daily activities or revolutionary alterations in thinking. Rather than add to our daily burden of things to do, they suggest modifications to what we already do. They are based on the principle that change, in a democratic society, requires the action of a fairly large number of people. The simpler the activity promoting change, the more people will participate. The higher the threshold (the perceived personal cost) for involvement, the fewer who will bother. In this book, the threshold to be crossed is one of understanding rather than of extra work.

My predilection for low barriers to change is based on personal experience. In 1970, during that cusp between '60s peace activism and the '70s environmental movement, my family joined a food cooperative. The Ecology Food Co-op was housed in a church basement, downstairs from a cooperative day-care center. Members were required to work a certain number of hours each week to sustain the enterprise.

Our first work assignment was to sweep, scrub, and straighten up the store after hours. We arrived for our shift with our toddler and were joined by another couple with a young child. The work took several hours, split among the four adults who washed and carried and stopped occasionally to attend to the children. Coming on the heels of our regular work day, the extra hours of labor in the damp basement—for the sake of being able to purchase organic peanut butter and lettuce wilted by storage in an aged refrigerator—seemed a kind of punishment.

Afterward, instead of driving home to a wholesome meal of co-op goods, we headed with our cranky toddler to a local fast-food outlet for hamburgers and fries. Our work for the Ecology Food Co-op had made us too tired, too late, and too discouraged to go home and chop vegetables. Rather, we opted for greasy, empty-calorie-laden, but supremely convenient prepared food.

As we sat down, we discovered the other half of our co-op work crew wolfing down their own fast food. Everyone ate in silence, staring at their plastic trays, humiliated at having been found in flagrante delicto, in mutual ecological sin.

We soon dropped out of the co-op, largely due to my stubborn refusal to work all day and chug home on a crowded commuter train to perform further work. The extra effort did not seem worth the reward. The food co-op, however, survived a good many years as the private buying club for a small number of dedicated people who did not mind the height of the hurdle.

Thinking back to that experience of almost twenty years ago, I realize that while the co-op movement struggled along due to lack of participation, the broader-based movement toward "natural foods" successfully transformed food merchandising. Today probably 75 percent of what the co-op had to special-order in huge quantities from great distances is readily available in supermarkets across the country. I can find bulgur, tofu, jicama, and organic soy sauce without leaving the confines of my local store.

American food retailing is amenable to change. The wide variety of goods now displayed on grocery-store shelves reflects our eating habits and overall values about food. In other words, we consumers have a very big say in what gets sold to us, although retailers do not go out of their way to point this out.

The potential to have an impact on what is sold at our local supermarkets and neighborhood stores is the basis for this book. The pro-biodiversity actions recommended here step up, chapter to chapter, from requiring minimal to more involvement. The book as a whole is modeled on the natural world around us: change evolves through tiny, incremental events that occur consistently and repeatedly over time. We need only modify slightly how we shop, what we eat, and what we plant, and we can change the world.

GENES, HYBRIDS, AND THE LOSS OF BIODIVERSITY

We live in a world that is incredibly and beautifully diverse. Glance out your window. In your backyard you might see dozens of different tree types and shapes, every possible color of flower, shrubs and bushes of many heights, perhaps a small tilled plot with red, green, and yellow fruits and vegetables hanging on bushes and vines. In a city street you might see people, dogs, cats, and birds of all sizes, shapes, and colors. Unseen at a glance is the multitude of bugs, spiders, and worms that fly, creep, and burrow across our planet, or the astounding numbers of microscopic plants, animals, and bacteria that live in the soil and under water.

We are so used to this diversity that we often overlook its significance: it is the most powerful biological fact of our planet.

What is the value of this complexity? Why is the world

around us so elaborate? Because nature needs a large pool of genes to select from as she reweaves genetic combinations. Options are critical in an unpredictable world. In geologic terms, glaciers come and go; oceans, lakes, and rivers appear and disappear; and the climate fluctuates. Even in the short term, habitats change radically when dams are built, pastures paved, forests stripped, and deserts overrun by dirt bikes. Genetic options provide the flexibility that ensures life. The organisms that adapt to change survive; those that do not, become extinct.

Adaptation hinges on the successful reworking of available genetic materials; it is a time-dependent, slow process. The gradual nature of genetic change is one of the significant causes for concern about global warming, the greenhouse effect. For if environmental changes outpace genetic remixing processes, extinction becomes probable for the affected species. Only a diversity of genetic traits, housed in a diversity of species, ensures that some will continue to live—for at least the short run. Those species adequately equipped with genetic remixing materials will survive to pass their "robust" genes on to their progeny. Those without a wide selection, broad repertoire, and varied genes will not survive, will not be able to reproduce, or will not reproduce much. Genetic uniformity, then, works against species survival. The species that hedge their bets by fielding the most diverse set of individuals, who embody the most diverse set of genetic instructions, are most likely to adapt—and to prosper.

Safety in Numbers

We can speak only theoretically of how an individual's or individual species' genetic diversity provides a secure foundation for its long-term survival. Realistically, communities of biological entities—plants and animals—survive by virtue of shared, intertwined diversity. When overall biological diversity is low, extinction can occur regardless of the richness of any one species' genetic diversity.

From the tropics to the tundra, even a single seemingly insignificant change can set off a biological chain reaction. The disappearance of a particular bird or insect essential for disseminating plant seeds or cross-pollinating plant blossoms can reduce the numbers of weeds, shrubs, or trees in a habitat, in turn eliminating necessary food or shelter for other species of animals. Biological diversity therefore cannot be viewed on a genetic or species level alone, because few if any plants or animals exist in the world without some important connection to living things around them. If we make the world inhospitable for one living thing, we cannot know for sure what other living things we will be killing off at the same time.

While late-night comedians profit from poking fun at people who "fuss" over snail darters or spotted owls, the protection of such creatures is much more serious than the one-liners reveal. Each time we subdivide an orchard or irrigate desert land we risk snapping critical links in biological chains. Once we have ensured the extinction of one form of life and belatedly discovered that we are responsible for an even greater loss, it is too late: what is extinct is gone forever.

Obviously, nature is quite resilient as well. If this were not the case, life on this planet would have ceased long ago. The first extinctions would have had an irreversible domino effect. It is not extinction itself that is so threatening, but the rate at which it occurs. Scientists have estimated that the rate of bird and mammal extinction between 1600 and 1975 rose between five and fifty times from previous rates. The rate at the end of the twentieth century is projected to rise between forty and four hundred times what scientists consider normal.[3]

Nature has a lovely, stately pace, a pace that works best for growing, building, dying, and the other natural processes that occur millions of times all over our planet every day. Examples abound: touring Yellowstone Park after a huge fire, one marvels at the elaborate, interlocked growth processes that spring to action, sometimes within weeks of a seemingly devastating disaster. Nature abides such cataclysms. The forest recovers—it can handle a fire in its own way, at its own pace.

But bulldozers, pavement, sudden changes in the composition of the air or water or rain—these human-made onslaughts frequently overwhelm the ability of biology to accommodate, adjust, and survive. The stripping of the world's rain forests, the changed climate around our cities, the worldwide trend to desertification—these are just a few examples of human-induced change that destroys the very fabric of the natural system. If we keep this up, we are going to be in for some horrendous calamities, perhaps more quickly than now imagined.

How many species can a given ecosystem lose before it no longer works as a system? We do not know. But biologists Anne and Paul Ehrlich have invented a compelling analogy based on an aircraft held together by many thousands of rivets.[4] Flying along in this craft, we could safely knock out one of the rivets. Indeed, if as we fly we keep poking out rivets one by one, we would be able to go on safely for some time because there are a great number of rivets, and airplanes are very strong. Still, common sense tells us that eventually we would poke out just one more and the craft would come apart. There is no way to know which of the rivets will be the pivotal one—until we have poked it out and it is too late.

The Ehrlichs warn us to quit pushing before it *is* too late, before the system that sustains our lives, and the lives of all other living things, loses its capability to carry us.

Biodiversity and Food

The earliest uses of physics-based power were centered around food—water wheels to grind grain, for example, or sailboats to trade food and spices. Even today, much of our interaction with the natural world has to do with getting food, as we clear grasslands and forests for crops or animals, dam and divert river and lake waters for irrigation, and terrace mountains into cropland. Every person needs to eat in order to live. The fortunate among us eat every day. So no matter how urbanized we may be, we maintain immediate and direct contact with the natural, biological processes of the world through our daily meals. While there may be little that a person feels able to change in distant rain forests or high-mountain dam sites, we can exploit our direct connection with biology via how and what we eat and where we get our food.

We have long been admonished to eat a mixed and varied diet. But the reality is that 50 percent of the calories consumed around the world every day come from just three crops: corn, wheat, and rice. Combine this uniformity of diet with the fact that most of these crops are grown from just a few plant varieties and you have the potential for a severe, potentially catastrophic narrowing of the global edible-plant gene pool. For example, in the United States, six varieties of corn account for 71 percent of the yearly crop, ten varieties of wheat dominate the harvest, and four varieties of rice make up two-thirds of the planted acreage. In Japan, two-thirds of local rice varieties have been lost in this century. India had 30,000 varieties of rice fifty years ago but today depends on just ten strains.

Most, if not all, of the varieties in popular use today are inbred or hybrid varieties. Indeed, the food-crop biodiversity crisis is largely rooted in the use of nontraditional plants. In scientific terms, inbred varieties (whether plants or animals) result from generations of crossing closely related individuals, producing offspring that are practically uniform genetically. Hybrid varieties are the offspring of genetically dissimilar parents. Typically, the parents of hybrids have been inbred for a few specific traits; the offspring manifest the carefully chosen traits of the parents.

It is not that inbred or hybrid plants are intrinsically all bad; in fact, they often grow faster, yield more, and harvest more easily than traditional varieties. That is why they are so popular. But they are carefully conceived for these characteristics, and they have a narrower, less flexible genetic base as a result. While they may yield well, to do so they usually need more fertilizer, more water, and more protection from pests and diseases than do traditional plants. Because of these requirements, modern, "improved" plant types tax the resources of the planet.

Moreover, inbred and hybrid plants and animals are often the products of large corporations. As their virtues are extolled through national and international advertising, farmers are enticed away from planting traditional seeds. In some cases, big corporations control the local seed market to such an extent that farmers are forced, not cajoled, into switching to nontraditional varieties. If farmers have no stock of the older, more genetically diverse seed, once they stop growing it, it is in danger of becoming extinct. This is how, in just a few years, important plant varieties can be lost to the world forever.

Seed banks, botanical gardens, and wildlands preserves act to prevent this type of extinction from occurring. Seed banks in the United States and around the world store collections of seeds, ostensibly to preserve their diverse genes for the future, when they might prove useful in increasing plant productivity

What Is the Future Return
on a Seed Banked Today?

When plants are removed from frequent contact with field conditions, their evolution is put on hold, warns botanist Gary Nabhan in *Enduring Seeds* (San Francisco: North Point Press, 1989, p. 97).

"I can imagine a scenario in which the rare, valuable plants now sheltered in seed banks, once released, succumb to newly arisen problems that did not exist when they were put into storage. Consider the demand for salt-tolerant plants that will exist by the year 2089, when conventional crops will fail on much of the world's desert farmlands that have been salinized by poor quality groundwater. Let us imagine that at that time, a crop breeder searching for salt-tolerant plants discovers that a twentieth-century scientist once banked seeds of a wild bean relative that grew on saline playas of the Mexican coast. That wild bean population

or fending off disease. Botanical gardens and preserves maintain living collections of rare plants for the same purpose. For either system to work, the seeds must first be located and collected, then handled in a way that protects their genetic structure and viability.

Neither of these preservation techniques are fail-safe, and their implementation suffers from lack of funds. Untold thousands of plants have already been lost forever, and more will go is now extinct in the wild, and no other sources of salinity tolerance are known for cultivated beans. He summons this sole accession up from the National Liquid Nitrogen Pool, where it has been stored for one hundred years. It arrives in the mail the next day.

"The scientist sprouts it in a peat pot in the greenhouse where he normally hybridizes his beans. Instantly, most of the plants become infected with a whitefly-transmitted virus strain that didn't even exist when the seeds were first stuffed into a tube and dunked in liquid nitrogen. The whitefly has evolved to resist the most frequently used greenhouse insecticides, and the virus itself has become more virulent as the greenhouse industry has expanded. Worried, the plant breeder removes the surviving plants from the greenhouse, and attempts to grow them outside in a small garden at his Eastern university. There, they fail too, overcome by air pollution, or by soil contamination from acid rain. The salt-tolerant genes have been rendered meaningless by the altered conditions they find on the ground a century later."

extinct before anyone has a chance to try to bank their seeds or otherwise save them.

The seed bank situation is particularly alarming. The United States has a National Plant Germplasm System in which seed samples held at regional repositories are backed up by duplicate accessions at the Fort Collins National Seed Storage Laboratory (NSSL)—at least that is how the system is supposed to work. Henry Shands, director of the national system, warns that that is only an ideal at this point. Shands reports that 25 to 35 percent of the system's seed samples are in danger and need to be grown out for a variety of reasons, such as to replenish seeds with low rates of germination and samples of inadequate size, or to verify that samples held at regional and national banks are indeed the same plant, or simply to produce the seeds necessary to provide for a back-up accession. Ongoing budget limitations ensure that it will take at least a dozen years to grow out the seed samples now at a crisis point, says Shands; any new accessions must queue up behind.[5]

A 1990 study by the National Academy of Sciences found the situation even graver. The academy reports that "a large proportion (almost 50 percent) of the accessions at NSSL are below the minimum desired size (550 seeds). Regeneration of these samples is urgently needed."[6]

A recent study conducted by the International Board for Plant Genetic Resources revealed peril at the international level, finding that a majority of the world's collected crop germplasm is not securely stored and some of it has been irretrievably lost due to financial and technical shortcomings. Seven of the seventeen "designated base" gene banks evaluated, including the NSSL, did not meet the board's registration standards. These seven represent exactly half of the storage space surveyed and 60 percent of the germplasm stored today.[7]

Botanist Gary Nabhan, author of *Enduring Seeds,* puts these reports in perspective with his reminder that of the

The End of the Nursery-rhyme Pig

Seed banks and plant conservancies, while admittedly of questionable effectiveness, at least exist. Few groups, however, are working to save animal breeds from extinction, defying the wisdom of the conserver's philosophy: don't throw those genes away; your grandchildren might want them.

Among the animals already extinct is one perhaps many of us knew in our childhood. The curly-coated Lincolnshire pig is the one in all the nursery-rhyme picture books, the common pig of the nineteenth century. A lean and hearty pig. Pig farmers in the twentieth century bred to satisfy consumers' growing taste for well-marbled, tender pork and bacon. They had no use for the Lincolnshire, and the breed became history in 1967.

Since then, of course, consumers have learned of the health risks associated with cholesterol and saturated fat and no longer want fatty chops and bacon. The loin of a Lincolnshire, were it available, might meet Heart Association standards today, but nobody thought to save the breed.

Scientists differ greatly in their estimates of the rate of loss of species: by the end of the century, somewhere between 10,000 and one million of the eight million species on Earth will no longer exist. Of the 175 known livestock varieties in North America, 80 are already considered threatened. Even cows, such as the Guernsey, that were common when we were children are on the "threatened" list.

160,000 crop samples delivered to U.S. Department of Agriculture plant introduction stations since 1898, only 5 to 10 percent are still alive and accessible. As a kind of "survival insurance, seed banks may be fine," says Nabhan, "but there will be tremendous losses if we assume that they are all we need."[8] Botanical gardens and preserves, too, are limited in what they can save, as well as by the biological fact that the long-term inbreeding that can occur within protected populations can itself destroy genetic diversity.

Why go to the effort of preserving ancient seeds and plants? Are we losing anything important—if the newer plants yield more and better?

Who knows, since currently insignificant traits might be priceless in the future. "For example," says Jack Doyle, author of *Altered Harvest*, "the genetic material found in one Ethiopian barley variety resistant to yellow barley mosaic now saves American farmers an estimated $150 million a year, and some Turkish wheat genes resistant to stripe rust have saved American wheat farmers $50 million annually since the 1960s."[9] Closer to home, Kent Whealy, head of the Seed Savers Exchange, reports that the stringlessness of green beans hinges on one gene that is now incorporated in 99 percent of the snap beans grown in this country.[10] If a disease attacked that gene, people would desire the older varieties. But would any be alive?

So who is encouraging the use of hybrid varieties? Consumers, because we demand plentiful, cheap beef, pork, and fowl, all of which are raised on volumes of hybrid grain, as well as unblemished produce of uniform shape and size. American farmers, because they prefer plants that ripen at the same time, hold up to mechanical harvesting, and produce bigger or weightier yields. And, most of all, seed companies, because inbred and hybrid plants are an entree to an expanded market for fertilizers and pesticides, and because they ensure annual seed sales, as the seeds must be recreated anew each year and are patentable. Not surprisingly, these companies are also typically heavily invested in biotechnology, a technology that makes a new spectrum of patentable plants possible—at unnatural speed.

Is a Perfect Ear of Corn Desirable?

Corn is genetically "the most tinkered-with plant in existence," says Margaret Visser, author of *Much Depends On Dinner* (New York: Grove Press, Inc., 1986, pp. 47–54).

"In 1893 a farmer called James Reid won a prize for his spectacular corn. He had inherited the strain from his father Robert, improved and perfected it for many years, and finally showed it at the Chicago World's Fair. It was called Reid's Yellow Dent, and it was the result of a failure and an accident. Back in 1847, Robert Reid's crop of reddish corn had failed in patches, so he planted the bald spots in his fields with a flint corn called Little Yellow. One type fertilized the other, and

Reid's Yellow Dent was the prize-winning result. The cobs were heavy with golden yellow kernels, all of them plump and evenly spaced from cob tip to beautifully rounded butt. Reid's Yellow Dent was snatched up and planted by whoever could get it; it spread across the United States. . . . Reid's Yellow Dent was so good, so obviously preferable, that other corns were forgotten, simply not planted.

". . . . The years between 1900 and 1920 in the United States and Canada were the years of the corn shows. . . . At the annual corn shows, farmers exhibited their best cobs and won prizes for the beauty of the ears. One of the great American myths was being created and accepted with ever-increasing confidence and fervour: that beauty—in corn as in anything else—is largely a matter of uniformity.

"In 1922 hybrid corn began to become commercially available. By 1950 the corn shows and the corn judges had been swept away: hybrid corns (based largely on germplasm from Reid's Yellow Dent) became available which were absolutely reliable, predictable, and capable of yields previously undreamed of. Farm machinery was invented, and once it appeared it had to be bought if a farmer wanted to stay in business. Corn genetics enabled farm machinery to operate, because it produced plants to order which the machines could handle.

". . . . In 1970 an epidemic of corn leaf blight struck the United States: oblong lesions appeared on the leaves, stalks weakened and fell, yield was greatly reduced. This new mutant strain of an ancient scourge victimized only one type of corn—but that type had been planted by almost every farmer in the country. This corn variety contained Texas cytoplasm. . . . Scientists hurriedly replaced the Texas with normal cytoplasm . . . and the danger was soon averted. This was possible only because alternative genes existed. One of the most popular traits of the labour-saving corn hybrids with Texas cytoplasm in their parentage had been the built-in resistance of these plants to corn leaf blight. . . . It took only a slight mutation in the blight to cancel out this scientific advance. . . . Yet even now, with the lesson of 1970 behind us, only six main strains make up the gene stock of nearly 50 percent of all corn grown in the United States."

A Link of Steely Strength

Agriculture is big business in this country. We not only feed our own quarter-billion citizens, but ship corn, wheat, and soybeans to the feed lots and cook pots of many nations around the world. Over the last century, the standard model in American agriculture switched from locally oriented family businesses to a huge industrial process in a conversion that, not incidentally, makes foreign marketing more achievable. One aspect of this evolution in agriculture is a trend toward monoculture—the growing of vast quantities of exactly the same thing.

The potato is a good example of this trend. In the United States, twelve varieties of one species of potato account for 85 percent of those marketed. But one variety, the Russet Burbank, garners the greatest acreage—about 40 percent in 1982.

Why? The Russet Burbank is the "McDonald's potato," the fast-food company's pick for its french fries. McDonald's demand is farmers' command in the United States. And as the company grows internationally, it spreads its dependence on the Russet Burbank to acreage overseas.[11]

While monoculture farming of inbred and hybrid plants and animals represents the fundamental element of the amalgamated link between diet and the loss of biological diversity, it is the meld of elements in the link that gives it its durable strength. Inbred and hybrid varieties require the use of pesticides and fertilizers, facilitate corporate control of farms and seeds, and provide a market for the techniques and products of biotechnology.

Why Poison Ourselves?

One of the sobering realities of the movement toward monoculture is the concomitant overuse of farm chemicals. The genetically uniform plants that are favored in modern agriculture are quite frail. While some of these plants show high yields, at least on paper, their successes depend on extensive outside help. The promise of miracle production from hybrid plants is fulfilled only if the plants are nurtured with fertilizers and protected by pesticides and herbicides. For example, since the introduction of hybrid corn in this country, farm use of fertilizer has increased tenfold.[12] Hybrid and inbred plants are like athletes who can accomplish prodigious feats of strength

Pesticide Tolerance

Some 600 chemicals used to formulate 50,000 pesticides are now marketed in the United States. The majority of these chemicals were registered for farm use before comprehensive safety testing and standards were initiated. Only 125 of the chemicals have undergone the now-prescribed health and safety testing necessary to establish a tolerance level—the level of application permitted on crops.

The U.S. Environmental Protection Agency (EPA)—the agency responsible for setting tolerance levels and for requalifying inadequately tested chemicals—ranks pesticide residues as the nation's number-three cancer risk. Pesticides are outranked by workplace chemicals and indoor radon gas. But the EPA's highly simplified calculations may obscure the full dangers of exposure of pesticides. The EPA assumes that just one chemical is applied to a crop, even though it is typical that several are applied. There are, for example, more than 100 pesticides registered for apples. And the agency overlooks cumulative or additive effects, despite warnings from scientists that the application of one pesticide atop another can multiply or alter the effects of both.

The following shows the human risk from dietary exposure to certain legal pesticide chemicals, in cancer cases per million people, as estimated by the National Academy of Sciences. These current estimates often exceed the Environmental Protection Agency's "acceptable" rate of one case per million; the EPA's tolerances, or maximum allowable residues, haven't been revised for these chemicals.

ACTIVE INGREDIENT	RISK	YEAR TOLERANCE WAS SET	MAJOR CROP USES
Captafol	594	1959	Apples, cherries, tomatoes
Captan	474	1955	Almonds, apples, peaches, seeds
Maneb	442	1957	Fruits, small grains, vegetables
Mancozeb	338	1962	Fruits, small grains, vegetables
Chlorothalonil	237	1961	Fruits, peanuts, vegetables
Benomyl	113	1972	Citrus fruits, rice, soybeans
O-Phenylphenol	99.9	1955	Citrus and orchard fruits
Acephate	37.3	1972	Citrus fruits
Oxadiazon	12.1	1977	Rice
Pronamide	7.77	1972	Lettuce
Glyphosate	0.27	1976	Hays, orchard fruits
Fosetyl Al	0.03	1983	Pineapples

Note: Risk estimates assume pesticide residues are at the tolerance levels, that 100 percent of acres with appropriate crops are treated, and that exposure occurs over a seventy-year lifetime.
Source: Sonia L. Nazario, "EPA Under Fire for Pesticide Standards," *Wall Street Journal,* February 17, 1989.

and endurance—but only as long as they have access to steroids, pain killers, supplements, and other ultimately nonsustainable, self-destructive crutches.

Since our agricultural system has accustomed itself to high-yielding, frail plants, it has become inured to the various props these crops need for survival. Their chemical dependency was first revealed to the world through the tragedy of the Green Revolution, when newly developed, high-yielding varieties of wheat and rice were dispatched to India, Mexico, the Philippines, and many other hungry nations across the world. The new varieties produced well—when given enough nitrogen fertilizer. But when the energy crisis of the 1970s took effect, the cost increases in fertilizer put the newly dependent small farmers out of business. In India, the cost of fertilizer imports rose more than 600 percent between the late 1960s and 1980. At the same time, imports of pesticides rose, and energy-dependent irrigation wells (suddenly mandatory for farming) were drilled around the country. As Doyle points out, "Ironically, in displacing traditional agriculture, the Green Revolution also displaced the very genetic variability that once worked to resist crop disease and insect infestation. Without pesticides, the early Green Revolution crop varieties—although high-yielding under favorable conditions—were often vulnerable to considerable insect damage."[13]

The Green Revolution will one day be seen for what it really was: not so much a new system of agriculture as much as a global tool for redistribution of control of agriculture—a kind of retrograde land reform.

In addition to chemical coddling, many modernized plants also require huge amounts of water. They produce watery-tasting, heavier crops (their water weight is a commercial advantage at the cost of nutritive value), put unnatural pressure on water resources in farming areas, and increase the energy demands of farming because of the need for pumped irrigation. Further, the fertilizers and pesticides used in the fields ensure that farm runoff is polluted. Through agricultural runoff, the chemicals that have been spread over fields and crops move into the nonfarm environment—into wildlife habitats and human water supplies.

The chemicals needed to compensate for the plants' weaknesses include huge aggregates of pesticides, fertilizers, and cosmetic/marketing enhancers such as Alar, the apple color-brightening chemical. The pesticide family of chemicals includes herbicides, insecticides, fungicides, nematicides, and rodenticides. Of these, herbicides are the most copiously applied: approximately 625 million pounds of herbicides were used in the United States in 1981, and the amount is growing exponentially.[14]

Agrichemicals add directly to the costs of farming and to the costs of produce. Worse, agrichemicals are often poisonous to other plants and creatures in the environment, including people; their indirect costs in damage to human health have not yet been fully calculated.

It is possible, of course, that not all agrichemicals are harmful in the quantities now used, but few have been tested adequately enough to allow conclusions on their potential toxicity. According to the Natural Resources Defense Council, "At least fifteen products that researchers previously believed had the properties of chemically breaking down rapidly and of being nonpersistent have been detected recently in surface- and groundwater across the United States. The Environmental Protection Agency (EPA) has placed several herbicides under special toxicology reviews, after experiments on laboratory animals showed them capable of inducing birth defects and cancer."[15]

So while environmental groups and chemical companies battle each other in costly, time-consuming litigation for proof of chemical safety, millions of pounds of compounds with unknown effects are being spewed onto our food crops. The fact that future research or regulation may cause withdrawal of some of today's popular farm chemicals cannot ameliorate the damage that may already have been done to human and other natural resources.

Are there alternatives that protect biodiversity? Certainly. Sustainable organic agriculture is viable, profitable, and represents an alternative that has kept humanity going for millennia. Even the National Academy of Sciences, in its recent *Alternative Agriculture* report, says that it works. And plants that perform well without resource-depleting props are hardy by virtue of their genetic diversity. Unlike hybrids, the plants typically used in sustainable farming are traditionals that maintain the breadth of genetic resources to resist a variety of pressures—from disease to pestilence to erratic weather conditions. A switch to sustainable agriculture would greatly broaden food-crop genetic resources because farmers would select the plants best adapted to their local conditions, taking into account soil characteristics, water supply, length of growing season, average and extreme temperatures, and number of daylight hours.[16] In this way, sustainable agriculture practices directly fortify the planet's genetic resources.

He Who Pays the Piper

The ownership of farms and agriculture-related businesses has been greatly consolidated in the United States and the world in recent decades. Although there were 2.4 million farms operating in the United States in 1985, just 7 percent of them controlled 56 percent of the nation's agricultural production, while fifteen agribusiness corporations sold 60 percent of all farm supplies, and about sixty companies performed 70 percent of the nation's food processing.[17]

Consolidation somewhat illustrates the connection between corporate control of agriculture and a narrowing of the genetic base. More eloquent still is the roster of major seed companies: the top ten seed-selling companies in the world in 1986 were the corporations Cargill, Ciba-Geigy, Dekalb-Pfizer, Lubrizol, Pioneer Hi-Bred, Royal Dutch/Shell, Sandoz, Suiker Unie, Upjohn, and Volvo. All are multinational; the majority are chemical companies. Only Suiker Unie and Cargill are agribusiness firms; only Pioneer Hi-Bred is solely a seed company. Surprised to find such heavyweights involved in selling seeds? Tiny seeds are big business: the world market

Exports and Acreage

There is a noteworthy ripple effect to the seeds planted in American soil, as agricultural historian James Wessel in *Trading the Future* (San Francisco: Institute for Food and Development Policy, 1983, p. 54, 159) reveals:

"In one decade, the top three export crops—corn, wheat, and soybeans—grew from one half to cover two thirds of our harvested cropland. Not only have export markets tended to narrow farmers' crop options, they have encouraged farmers to accelerate a shift from mixed livestock and crop production toward specialization in corn and soybeans, especially in the North Central states.

"According to agricultural economist Philip M. Raup, we are 'creating a pattern of one-crop, export-based agricultural activity in the corn, soybean, wheat, and sorghum regions that is very similar to the type of monocultural dependence formerly associated with colonialism. In an important and sobering sense, the grain belt of America is acquiring the characteristic of a colony.' Nothing more characterizes the economy of a colony than the

for crop seed is $45 billion annually; the U.S. market is $4 billion.[18]

The result of multinational ownership of seed supplies is restriction of the types of seed available worldwide. Consider that:

· More than 48 percent of all nonhybrid seeds are available from only one source.

boom and bust linked to fluctuating world market prices for primary exports. Also, in a colonized economy the rewards of production are captured primarily by the largest producers, as well as by the commercial interests off the farm.

"When farmers produce a variety of crops, or both livestock and crops, they have some built-in protection against losses when prices decline. If feed grains hit a price slump, farmers can always feed part of the crop to their livestock. And when farmers produce a variety of crops, they are somewhat protected if prices fall for one. With greater crop specialization, such back-up protection is lost.

".... It has been no coincidence that the countries experiencing the greatest growth in U.S. food processing and retailing subsidiaries are the same countries that received the bulk of U.S. farm exports in the two decades prior to the export boom of the 1970s.

"U.S. marketing and promotional strategies were transplanted abroad as well, helping to create new consumer demand for a 'global supermarket' on the American model—one that emphasized a highly processed grain and meat-centered diet, for those who could afford it."

- A single corporation controls 66 percent of the world's banana germplasm.
- Two companies control 43 percent of all barley seed types.
- Four companies control 79 percent of all bean varieties.
- Six companies control 66 percent of all lettuce types.

With such control, these corporations can afford to play the market the way that suits them best—through uniformity.

Corporations thrive on it. Economy of scale is a basic tenet of the modern industrial system. Diversity in a product line is costly—in terms of planning, processing, packaging, advertising, and so forth. The now-leading multinational seed vendors established themselves by buying out family-owned seed companies and replacing what were multiple, regionally adapted seed collections with collections of a few (more profitable) hybrids and patented varieties. Why should they go to the expense and bother of selling five kinds of peas (specialized producers for various climates around the country) if, by their market dominance, they can offer just one patented variety that will produce "well enough" in all regions?

Vertical integration into agriculture is another card these companies play to strengthen their stance in the marketplace. Such integration enables them to entrench their primary products in a marketing system that reaches farmers with everything from seeds to plastic pipe and from herbicides to machinery. Since they produce farm chemicals, they have a powerful incentive to develop and promote agricultural practices that involve intensive use of the products from which they profit so greatly. They are powerful proponents of the monoculture crops that require chemical nurturing.

Designer Genes

Biotechnology, the new science that permits laboratory manipulation and recombination of plant or animal genes, is the twentieth-century tool for fine-tuning the industrialization of agriculture. If, as stated earlier, uniformity and control are major goals of agribusiness, biotechnology is a key mechanism for achieving those goals. Instead of years of gently guiding nature toward a selection of desirable characteristics, biotechnology uses recombinant DNA, cell fusion, and other techniques to rapidly change the genetic makeup of living organisms. By recombining the genetic substance of various living things, biotechnology can provide exactly uniform plants or animals that meet industry specifications.

Such genetic manipulation moves us considerably closer to viewing living organisms as industrial commodities ripe for control. As biotechnology watchdog Jeremy Rifkin notes, "Whoever controls germ plasm, and therefore genetic engineering, is as important as who controls oil. With the big chemical companies moving in to collect seeds and patent animal embryos, they potentially will have power over many of the living things of the future of this planet."[19]

Those involved in biotechnology promise that gene splicing, tissue culture screening, and protoplast fusion offer cures for the shortcomings of traditional breeding methods and a means for rapid development of heroic, high-yielding plant varieties resistant to disease, insects, climatic factors, and herbicides. They suggest that because current agricultural practices accelerate the evolution of new pest types, only biotechnology can respond quickly enough to produce new cultivars resistant to infestation. Because technology has sped up evolution, biotechnology will come to the rescue by speeding it up even further.

Navajo-Churro Sheep

What is true for modern plant varieties is true for "improved" animal breeds, too.

Sheep rearing is common in the Northern California valley in which I live and work. The newer breeds preferred by most farmers in the valley are high yield, whether they are raised for meat or wool production. At my workplace, the C. S. Fund, we raise Navajo-Churro sheep, an older breed that gives less meat and wool per animal than our neighbors'. I think it is debatable, however, which sheep truly yield more or better.

Since the modern sheep were bred for one goal—production—their genetic makeup is focused toward that end. In contrast, our more primitive sheep have very diverse genes, the result of hundreds of years of husbandry by native people and Chicanos in the Southwest. Our sheep, therefore, retain many qualities not found in the flocks up the road. What could make sheep more valuable than lots of flesh and fleece? Well, our sheep are excellent

Indeed, with gene splicing, inhibitive natural boundaries are broken down. In the laboratory, it is now possible to move genes from any one organism in the plant kingdom to any other, cleverly outmaneuvering Mother Nature.

In the past, one natural limit to the quantity of chemicals that could be sprayed or otherwise applied to food crops was mothers: the rates of lamb survival and twinning are high. These animals also eat almost anything and thrive on the sparse coastal pasturage we provide them, without need for expensive, pesticide-riddled grain or hay hauled in from out of the county. They flock well and even attempt to defend themselves by collectively attacking intruding animals, so loss from predation is low. They are resistant to hoof rot, a common and expensive problem in wet, low-lying areas. In fact, we rarely have need for a veterinarian.

In contrast, the other sheep in the valley, bred out onto a genetic limb, demonstrate a generally lower birth rate and lesser ability to care for their offspring. They don't flock well or act in self-defense, and they get sick a lot—a problem owners try to forestall with preventive medicines.

The modern valley sheep that make it to market do indeed weigh more than our Navajo-Churros. But by nature, without props or assistance, our sheep are the much better producers and survivors. We think we have the high-yielding animals, once the appropriate factors are considered.

the ability of the chemicals to harm the plants as they worked their effects on insects or pathogens. Biotechnology, however, enables scientists to tailor plants to have a specific relationship to agrichemicals, potentially producing herbicide-tolerant or pesticide-tolerant varieties. Predictably, the biotechnology scientists making new plants that are tolerant or virtually immune to agrichemicals are funded or employed by the companies that produce the chemicals: greater plant resistance means greater chemical sales.

In 1985, Monsanto allocated $104 million for biotechnology research and development (20 percent of its total research budget).[20] Monsanto's interest is driven by ownership of four seed company subsidiaries and investments in four major biotechnology firms. While annual sales of herbicides amount to roughly $4.5 billion worldwide, Monsanto's sales were $1 billion in 1982 and likely exceed that now.[21] For Monsanto, it is a market worth the expense of experimentation.

If plants can be engineered to survive the application of greatly increased volumes of chemicals, then the industrial farmer can spray without hesitation or concern—and expose agricultural workers, nearby plants and animals, the water supply, and consumers, ultimately, to greater quantities of toxic substances. Other less obvious dangers of biotechnology exist as well: for example, once a plant is created that is herbicide-resistant, what is to stop it from crossing with a noxious plant or weed? How would the weed be eliminated then? How would the genetically conferred resistance be controlled once the organism is loose in the environment?

Scientists assure us that they have safeguards against such accidents, and they may well be right. But even scientists can be misled: Challenger, Chernobyl, and Bhopal are the most memorable recent examples of dreadful tragedies that were in each case preceded by soothing assurances from honest and well-intentioned experts. Biotechnology backers ask us to take environmental risks without being able to demonstrate their ability to deal with all the potential results. They ask us to tolerate the release of genetically engineered organisms—for example, the ice-minus bacteria known as Frost-Ban—when they cannot predict what effect these bacteria will have on other organisms in nature, or what might happen if they begin to reproduce on their own. The purpose of ice-minus microbes is to protect certain agricultural plants from frost damage. The point is not so much what might happen in a small-scale plot, but the unknowns of application to thousands of farm acres across the country. If the unexpected occurred, how would the organisms be contained?

This is not to say that all biotechnology is bad or that we should not use modern techniques to improve our agricultural system. But we must ensure that decisions on new technology reflect the desires of those who would likely bear any ill consequences, rather than solely the power of those who stand to profit.

Let us explore the notion that biotechnology is also a *solution* to the loss of biodiversity: through a recombination of genes, biotechnology might expand biodiversity. While at first blush this seems a reasonable idea, remember who is doing the research. The products of biotechnology are patented, typically corporate products—costly and of limited accessibility. Biotechnology represents the wrong direction in thinking about solutions.

I recall sitting at lunch a few years ago with an earnest scientist of genetic engineering. As I raised questions about the rapid decline in genetic diversity, he reached over to pat my arm, to soothe me, saying that in the next decade, at most, science would marry enough microbiology to computer science to create whatever genetic combinations might be needed.

I was not soothed. This is cataclysmically dangerous thinking, aside from being untrue. Even if it were possible to create entirely new organisms through manipulation of genetic material—and this seems a technical impossibility for at least another ten years—it would never be acted upon, because the required ultra-high-tech DNA work is phenomenally, almost cosmically, expensive. Even our wealthy American society does not have the material resources to spend what it would take to recreate what has been lost: the thousands of rice varieties, the tens of thousands of wheat varieties, the curly-coated Lincolnshire pig, the Norfolk Trotter horse. Even if science could recreate lost life—a perilously arrogant proposition—there would never be enough money to implement the plan.

Even more important, this viewpoint perceives the world as made up of various independent units, rather than as a highly interwoven system. More than saving various kinds of South American potatoes or cheetahs, we need to save the complex biowebs that sustain them and that they, in turn, sustain. The tendency of some to examine each living thing *ex situ* permits all kinds of wrong-headed thinking. Biology works in huge, beautifully complicated ecosystems, not as dissected components lying cold on stainless-steel laboratory tables.

Our biodiversity crisis can be abated if we recognize that

our job is to support nature in her own sure and proven methods of healing; if we stop poisoning, clear-cutting, and mining the earth as if it were an ever-resilient, limitless pit of jewels. If we permit our adeptness at science to delude us into thinking that we can outwit and outdistance nature, we will end up overpowering ourselves—to our everlasting shame and detriment.

Who's Guarding the Cattle Pen?

One of the most dramatic and least well attended areas of genetic decline is in livestock breeds. Changes here are typically known only to industry. While we consumers are generally aware of the variety of apple we are buying (is it Granny Smith or Delicious?), few of us know what breed of cow our milk comes from, much less what breed of chicken we sautéed for dinner last night.

Livestock experts project that in the next few decades we will see a 45 percent loss in North American livestock species. Of the 175 known breeds of livestock in Canada and the United States, 80 are listed by the American Minor Breeds Conservancy (AMBC) as declining or rare breeds.

And, unfortunately, the livestock extinction crisis is not specific to North America. Around the world, different countries are experimenting with a variety of livestock conservation strategies, many of which are dependent on nongovernmental organizations (NGOs). The best examples include:

- Great Britain, which began to formulate a NGO Rare Breeds Survival Trust more than a decade ago and now has a network of private farm parks that have stabilized the populations of most of the country's remaining breeds.

- Brazil, which, in a government-assisted program that is based on a United Nations Food and Agriculture Organization recommendation, is attempting live-animal or cryogenic preservation of all the identified rare breeds in the country.

- France, where funds from commercial cattle registries are used to employ people to track down herds of old-type, rare cattle and take semen collections for cryogenic storage. Other breed registries have not followed suit thus far.

- Switzerland, where a ten-year-old NGO has bought up all the rare breed stock it could find, except for cattle. The organization "rents" the animals to farmers with the understanding that the animals must be carefully bred to increase their numbers and may not be slaughtered without permission.

In the United States, the AMBC is the only organization dedicated to the conservation of livestock breeds. It is a grassroots organization; no U.S. government agency is safeguarding animal germplasm for future use. The AMBC has compiled a livestock census but does not have the resources to own animal populations.

EVERYONE SHOPS,
EVERYONE EATS

Supermarkets provide a cornucopia of foods—and represent a storehouse of opportunities for consumer action to benefit the planet's ecosystem. Almost every commodity based on living things is sold in American supermarkets, from herbs and pharmaceuticals to hand cream and fresh ears of corn. Food shopping puts many of us more closely in touch with the world of living things than any other aspect of our lives. This chapter shows how we can achieve more than just the acquisition of another week's groceries with each trip to the food store.

Since supermarkets do not sell MX missiles or drums of benzene, our shopping—a demonstration of our economic clout—cannot directly address problems of nuclear war or environmental pollution. But while squeezing tomatoes or

Cartons of Eggs and Milk

Nestling brown eggs, instead of white, into your shopping cart is a simple way to nurture the planet's genetic resources. Once you get past the shell, brown eggs do not differ significantly from white. But white eggs represent a genetic disaster in the making: more than 90 percent of the three billion eggs sold in the United States each year come from one kind of chicken. In fact, virtually all commercial egg-laying chickens in the United States come from just nine hatchery sources. What would happen to our egg supply if some pestilence wiped out that breed?

One thing is for sure. The white Leghorns that lay our white eggs cannot lay brown eggs. If even a moderate percentage of shoppers switched to brown eggs, the market would respond with a shift in our national egg-producing gene pool from one breed to two. Two breeds is still a very narrow genetic pool, especially given the large number of poultry breeds available, but it is a good start. Would you rather sail on a ship with one lifeboat, or two?

Milk presents a similar opportunity. Since 95 percent of our milk comes from just one kind of cow, the Holstein-Freisian, we can strike another small blow for gene-pool conservation by buying milk produced by other cows, such as Jerseys or Guernseys. Your family will never taste the difference. In some communities, milk origin is revealed in labeling. Otherwise, stop at a family-dairy store to see what is offered, or call a local milk distributor. You can find distributors listed under "Milk" in your telephone directory.

thumping cantaloupes in the produce section, we are in a position to take actions that can directly affect the composition of the world's gene pool. As we walk the aisles, tossing our selections into our cart, we are casting our votes in a continuing referendum on what should be sold. Food-store owners stock what they believe consumers want, and they gather their data from cash-register receipts. Consumers' demands are passed to farmers through retailers' orders. What retailers order is what farmers grow. And what farmers grow, as shown earlier in this book, is a major component of the biological diversity crisis.

While it may seem that no simple change in shopping habits could make a difference worth troubling about, history proves otherwise.

Think for a minute about the international Nestlé boycott. At its inception in 1974, the purpose of the campaign was to somehow pressure purveyors of infant formulas to stop marketing their products in poor countries in ways that resulted in infant deaths. Campaign organizers singled out the Nestlé corporation to serve as an example and target for the boycott. Initially, they were daunted to learn that not only is Nestlé the world's largest food corporation, it is also the most ubiquitous: Nestlé products are sold in almost every country of the world. As the boycott took hold, it was Nestlé's very size and ubiquity that made it an ideal target. Nestlé's omnipresence—products in nearly every store in every nation—provided the greatest number of opportunities for people around the world to participate in the boycott: an ideally low threshold for involvement. The application of widespread and continual pressure assured the boycott's success.

While redressing the loss of biological diversity cannot be boiled down simply to boycotting one or another corporation's goods, our supermarkets are prime sites for making choices that can constructively alter the planet's gene pool. We just need to know what to look for and what to buy.

Beauty Is Skin Deep

A blemish, by definition, is superficial: something that affects appearance but not function. My mother taught me this lesson many times during my childhood. When I would run to her with a banana or an apple with a "bad spot," she would take the imperfect fruit, bite off the blemish, and return the remainder to me: it was a "perfectly good piece of fruit." Eventually I learned that the brown spots were only of aesthetic concern. I was not taught to eat spoiled food, just food that was safe but cosmetically below par.

While no one wants to eat unappetizing food, our notions of how food should look have been driven to extremes by high-price food-industry marketing. American magazines are full of four-color food advertisements in which every tomato is brilliantly red, perfectly spherical, and blemish-free. Juicy hamburgers fairly burst from flawless rolls. Golden beer sparkles in frosty glasses. Ad agency specialists have a toolbox of tricks that ensure that the food portrayed looks not typical but ideal. Understandably, the goal of such advertising is to create consumer desire, and beauty is desirable—even if it is in produce.

But just as most of us do not look like movie stars or models (and the stars and models themselves show imperfections without benefit of makeup and controlled lighting), most tomatoes, hamburgers, and glasses of beer do not resemble their glossy-print renderings.

However, once smitten with an image of the ideal, we expect to be able to find something comparable in the grocery store—never mind the fact that the ad agency sorted through hundreds of tomatoes to find the right one. Marketers know that the closer the item on the shelf resembles the picture in our minds, the more they can sell and the more they can charge. So they pay the highest prices to producers who can emulate the ideal.

This pressure for food to appear a certain way influences farmers to reach for seed that provides uniform products as close to the ideal color, shape, and size as possible. Flavor and nutrition typically do not enter into their considerations, since American consumers make purchase decisions based on sight, feel, and (if we really know our cantaloupes) smell. So food is bred to *look* a certain way, as consistently as possible.

This system of selling food based on idealized representations deals a heavy blow to genetic diversity, because uniformity in crops is biologically and environmentally dangerous. The idealized, good-looking, blemish-free, midsize tomato comes from carefully bred seed. Never mind that these tomatoes have skin tough enough to rebuff an average kitchen knife and that their pulp has the flavor of wet cardboard. What is important is that these tomatoes withstand pesticides, mechanical harvesting and processing, and lengthy overland transport. If we demand that "model" look, these are the tomatoes that grocers will request and farmers will grow.

How can we combat the insidious trade-off that the food industry has made between food value and aesthetic perfection? The single most important thing we can do is to change our perception: blemish-free is not necessarily better. Taut, flawless skin may indicate the presence of unseen poisons. A firm feel may conceal an interior devoid of flavor. A perfect form may connote use of hybrid seed. In a recent survey, California shoppers were shown photographs of perfect oranges and blemished oranges: 78 percent said that they would not buy the blemished fruit. When told the blemishes indicated an absence of pesticides, 63 percent opted for the blemished fruit.[22] What do you do with a changed perception of what is good? Buy organic or off-grade produce.

Sometimes it can be tricky to discern what is organically grown, since the best organic produce looks just as unblemished and healthy as the chemically nurtured goods. In the average American supermarket, organic produce often simply is not available; if it is, it is clearly labeled. To find organic produce at smaller stores and roadside stands, inspect the produce packing box, or ask the workers who are replenishing supplies. If you do not see any organic produce, ask for it. Create a demand.

Locating off-grade produce means checking your regular

greengrocer's display for fruits or vegetables with greater vari-
ation in color, size, or shape, some blemishing, and lower
prices. These factors suggest produce that is more genetically
diverse, the product of nonhybrid plants. Depending on the
store, such an opportunity may crop up only occasionally,
when a regular supplier is sold out or another supplier offers a
bargain. Ask your greengrocer about the source and encourage him or her to seek nonhybrid produce. Even the most scientific and industrialized system of food production is subject
to the vagaries of nature (rain, heat, insects, disease), so a
good supermarket buyer always has alternate sources of supply. And his or her deviation from the normal system provides
us the opportunity to make a statement on the ecosystem's be-
half.

Choosing off-grade produce could add up to a cash savings
for you and a genetic-resources saving for the planet. Save
money while you retrain your eyes—and your family's—to
see imperfections as indication of healthier food. Aside from
helping to preserve the gene pool by providing a market for
nonconforming produce, we can learn to value diversity over
uniformity. As we become more accepting of differences in
our food, we are gradually more likely to see diversity as a sign
of environmental health and uniformity as an indication of
biological totalitarianism. After all, real life is imperfect.

Will our selections really register in an important way? Yes,
because food-store after-tax profit margins are tiny, sometimes as low as 2 or 3 percent overall. But the popularity of
fresh produce propels this department to prominence for re-

tailers: produce sales provide 21 percent of pretax profits at conventional groceries and 27 percent at larger stores.[23] Retailers are extremely sensitive to consumer choices. They know how much they ordered of a given product; they carefully monitor how much is sold; and, with laser-scanning at the register, they can get almost instantaneous feedback on consumer patterns.

If we choose the more genetically diverse produce—either organic or off-grade—the retailer will order more. If we do it consistently, farmers will get the message. If we create a demand for genetically diverse produce, the market will provide it. After all, supermarket owners are interested in profits; they do not have a vested interest in hybrids or pesticides.

Infiltrating the Menu

Many of us chat occasionally with the produce manager, one of the most visible of supermarket employees. This is a great opportunity to voice your produce preferences. As he or she splashes water on the lettuce and restacks the oranges, mention how much you appreciate the chance to buy organic produce, odd-size apples, and different kinds of onions. Make jokes about blemishes being only skin deep. If you are really bold, tell the produce manager how you would love to see something new: orange tomatoes, Japanese eggplant, fresh herbs. Produce sections do not have the slotting allowances (shelf space allocations) typically required for new products in other sections of the market, and that makes it easier for managers to try out unusual offerings. Of course, in many large supermarket chains, the produce manager has only limited authority because purchase decisions are made centrally for the various districts. But when some local produce buying is permitted, your comments can have a powerful effect.

It is easy to become habituated to wandering a set route in the produce department, from the cardboard tomatoes to the watery iceberg lettuce and on to the mushy apples. Retailers abet our habits by maintaining huge displays of these standard items. Learn to look carefully at the week's produce offerings; the unusual, local items are apt to come in limited quantities and will not be given prime floor space. Even when organic or off-grade fruits and vegetables are not available, you can still cast a vote for genetic diversity by trying something new.

While the produce section provides the most direct opportunity to "shop the gene pool," similar chances for supporting biodiversity exist in other parts of the store.

Give the store a cursory once-over. If you get beyond the packaging, it is easy to see that a great deal of what we eat comes from grains. The world over, 75 percent of human nutrition is based on just three grain plants: rice, wheat, and barley. The form may be whole grain or cereal, bread, pasta, crackers, or meat.

Yes, meat. In the United States, livestock consume ten times the amount of grain that we do. And for every sixteen pounds of grain they consume, they return to us one pound of fat-laden protein—a quite inefficient conversion of nutrients.

Beyond Wheat, Rice, and Russet

INSTEAD OF	CHOOSE
wheat bread	multigrain bread
	rye or pumpernickel bread
	barley bread
	oat bread
	cornbread
	sprouted-grain bread
	bagels and bialys
	tortillas (corn or flour)
	pita bread and lavosh
white rice	brown rice (long or short grain)
	whole wheat kernels
	barley kernels
	millet
	cracked wheat (bulgur)
	buckwheat groats (kasha)
	hominy
Russet potato	new potatoes
	Yellow Finn potatoes
	Yukon Gold potatoes
	blue potatoes
	sweet potatoes and yams
	taro
	winter squash (all kinds)
	plantains (baking banana)

We can contribute considerably to genetic diversity simply by eating a bit less meat. In doing so, we encourage farmers to plant a variety of grains for direct human consumption, rather than just feed grain for cattle, pigs, and fowl. A shift away from an animal-based diet could also help balance nutrient needs around the world by freeing resources and foodstuffs for export. After one experience of holding in my arms an African child who was dying from malnutrition and starvation, I have felt a powerful interest in seeing that the dietary choices I make do not discriminate against the poor and the needy. Babies should come before hamburgers.

GRAINS

How to increase the amount and diversity of grains in your diet? Eat the cuisines of other countries. Billions of Asians base their meals around rice; open up a Japanese, Chinese, Thai, or Korean cookbook for a wealth of ideas. East Indian cooks are masterful with wheat and rice—chapatis in the north and bryani in the south. In South America, corn is eaten fresh and dried, whole kernel and ground. Rediscover the European grain favorites of barley, rye, and buckwheat—when was the last time you treated yourself to buckwheat groats or buckwheat pancakes?

Pasta—that great human invention—deserves special mention. Standard Italian-type pasta is made from Durham, or hard, wheat. This type of wheat, different from the wheat used in making bread, is often grown as a winter wheat, in a cycle that alternates with the softer summer wheat. By eating both bread and pasta, you encourage production of different wheat varieties.

Other cultures are even more creative when it comes to making a wide assortment of pastas, which not incidentally keep for a long time and are an excellent way to preserve valuable nutrients. In Asian cuisine, pastas are made not only from wheat, but also from rice, mung bean, and buckwheat flours. In India, garbanzo bean flour is prepared into pasta.

LEGUMES

Speaking of garbanzo beans, legumes are wonder plants. They have the amazing characteristic of returning to the soil, through nitrogen fixing, important nutrients that other crops remove. If there is enough demand for beans that farmers can alternate crops, they can also use less chemical fertilizer in their soil. In addition, legumes are an excellent source of protein—especially when eaten together with grains. And they can be easily dried and stored for long periods of time with little loss of nutritional value. They are truly a food for the world.

In America, however, there is a negative connotation to beans; they have the image of a "lower class" food. For example, while asparagus or artichokes might be ordered as part of a formal, candlelight dinner, beans would be ordered only at a diner or roadside cafe.

One reason for this disparity between America and many other parts of the world is lack of diversity. When we say "beans," we think most readily of the small, white navy bean that's the stuff of pork-and-beans—hobo fare. Over the last ninety years, the varieties of beans grown in the United States collapsed from hundreds to just two or three primary types—in tandem with the growth of agribusiness-style farming. In New York State alone, where 260 kinds of common beans were once raised on small farms, only two kinds are now produced.[24] But many kinds of beans are still being grown, especially in other parts of the world: adzuki, fava, marrow, mung, orange lentil, yellow split-pea, and urad, to name but a few. Check your supermarket shelf.

Soybeans are the legume grown in greatest quantity worldwide. In the United States, they are grown predominantly for cattle feed. In other cultures, the soybean is revered for its nutrient content and versatility as a food. Most often the beans are cooked and pressed to produce soy milk or soy cheese (tofu), both primary ingredients in the cuisines of China, Japan, Thailand, Vietnam, Korea, and Indonesia. Tofu is typically stocked in the produce section of American supermarkets, to give it proximity to Oriental vegetables for stir-frying, but it has many possible uses. Tofu is just as well suited to extending scrambled eggs at breakfast as it is to stuffing pasta shells for dinner.

ETHNIC FOODS

A supermarket's ethnic-foods section is an excellent place to expand culinary horizons and nurture biodiversity. Winter months, when fresh produce is at a premium, are particularly good times to investigate foreign flavors. Dried mushrooms of Chinese or Italian origin, baby corn, dried chestnuts and water chestnuts, dried chilis and lotus root, pickled cactus and preserved plum, fermented black beans, gefilte fish, seaweed, and lychee fruit are all readily available to add new interest to meals.

If we support monoculture by limiting our choices in meal planning, then supermarkets and farmers will oblige us. While much of what is offered in the ethnic section is imported, not all of it is. And, if we create markets for ethnic foods, not only do we keep those products alive overseas, we could create a domestic interest in producing them. Even small increases in demand can prompt greater attention from retailers. For example, a while back I talked to a neighborhood group about biodiversity and brown eggs. A few months later, one of the group reported that soon after my talk, the few cartons of brown eggs routinely stocked at the local supermarket kept selling out very quickly. Within a few weeks, the supermarket realized that demand was up and responded by allocating more cooler space to brown eggs.

BEVERAGES

Soda pop, both diet and regular, is quickly replacing milk as the favorite American beverage. While fruits still figure in the appeal of many sodas (lime-flavored, cherry-flavored, and so forth), fruit and vegetable juices themselves are rarely consumed these days. However, not only do natural juices provide valuable nutrients (as compared with sweetly flavored water), they are usually free of chemicals and preservatives. Most important to biodiversity, these juices encourage production of a wider variety of fruits and vegetables because, for example, the best apples for juicing are not the best for eating, and vice versa.

In general, the task of the crusading shopper is to *slowly* change from the robotlike behavior of buying each week the same hard tomatoes, bland apples, and white eggs. Often when we think of changing how we shop and eat, we think about trying completely new foods. That is not a bad idea at all, but it may be more genetically significant to shift to variations of the foods that typically make up our diet—for example, shifting to orange tomatoes, a new kind of apple, and brown eggs.

By buying different types of traditional foods we expand the gene pool; buying several different lettuces or onions is direct resistance to monoculture and factory farming. And by selecting the less highly bred, more "primitive" foods, we provide farmers the incentive to cultivate plants that require less water and fewer chemicals.

A cautionary note. I once asked my wife to critique my cooking. She replied that I needed to reduce but one ingredient to be a good cook: enthusiasm. She was referring to my use of spices, but similarly, as I have discovered new foods, from black beans to Napa cabbage, I have sometimes assumed that my family would be intrigued by a week of unremitting experimentation.

As Shakespeare said, "Use all gently." Don't exhaust yourself and alienate those around you by plunging into a rabid campaign to alter the earth's gene pool in a few weeks. The goal of this chapter, and of this book, is not for you to become a revolutionary, but an opportunist: one who spots chances to do the small things that can broaden the gene pool—even just a little.

To Everything
There Is a Season

Nature has a four-phase cycle: birth, growth, reproduction, and death. Nature's bottom line is creation, but her production line is in a constant state of disorder: at any point in time, unknown numbers of organisms are at a different phase of the cycle. The hallmarks of nature's system are diversity and flux.

The business cycle, on the other hand, involves acquisition, transformation, and sales. Business's bottom line is profit, and its production line is carefully controlled so that at any point in time the right number of widgets are moving through to ensure a steady profit. Predictability and uniformity characterize successful business systems.

Attempts to adapt nature to the business cycle—to make her products predictable, uniform, and consistently available—contribute to the biodiversity crisis. When we pile melons and lettuce into our grocery cart in midwinter, or acorn squash and cabbage in midsummer, we encourage farmers and food retailers to override nature in deference to our gustatory desires. If we instead fill our kitchens with nature's seasonal provisions, we can steer the food industry toward greater respect for nature's cycle.

In the business world, production is readily mediated through control of raw materials, because plastic, steel, wood, and the like can be stockpiled without concern for spoilage. In food production and retailing, however, raw materials cannot so easily be stockpiled. Nature is the supplier—at least of fresh foods—and, if left to her own devices, she produces when and as she pleases. Sometimes she supplies too much, at other times too little or none at all.

The vagaries of nature are irksome to retailers, especially those who apply the business model to their operations. They are selling something that may or may not arrive at the time and in the quantity they desire, or that may be available in saleable form for only brief periods. But retailers and producers do not like to feel so out of control, and nature is typically blamed for this "problem." The solution has been to tinker with the supplier—nature—to bring her cycle into line with business's bottom line.

Commercial fruit producers, for example, favor varieties that ripen all at once and that can be stored a long time. Varieties that ripen over a period of months are troublesome: to operate a fruit business in tune with nature's proclivity for gradual ripening, producers must keep pickers, packers, and shippers on the payroll for months, a practice that curtails profits. To be sure, industrially oriented farming is monetarily rewarding for a few, but its more widespread biological and human consequences are hardly positive. Agribusiness farming can contribute to abusive labor practices. It can fail to deliver truly fresh produce to the consumer. And, worst of all, it leads to a dramatic narrowing of the gene pool.

The apple tree in my front yard represents one of the fruit varieties losing ground to the sanctity of profits. My tree is a Gravenstein, a traditional variety in my county that produces simply delectable fruit of unique coloring. These juicy, sweet apples do not keep for long, and so are not favored by commercial growers and retailers; but that poses no problem at my house. Since the fruit ripens gradually, we begin eating apples in late August (impatient children begin a bit earlier) and have a steady supply, more than enough for my family, until Thanksgiving.

The characteristics that make the Gravenstein work so well as a backyard apple make it undesirable in the commercial food industry, except for processed products such as applesauce and juice (and apples sold for processing bring lower prices than those sold for eating). In contrast to the Gravenstein, the ideal commercial eating apple ripens all at once, lasts without visible decay, and is cosmetically impervious to the low temperatures and gases used in storage facilities. If the big, shiny apple you bite into in January has a mealy texture and a flat taste, that is the price you pay for demanding an apple out of season.

In times past, prior to the Industrial Age and the idea of manipulating nature, people harvested when their crops were ready and ate what nature provided. Their diet changed seasonally as different plants came into fruition or as herds of animals migrated. Too, they devised ways of preserving nature's immediate bounty for future meals, through drying, salting, spicing, chilling, or canning. They made dietary changes to accommodate nature, rather than vice versa.

Today, science and technology make it possible to sidestep nature so that we can have more of what we want, when and as we want it—that is, everything, fresh, all year long. Accomplishing this, through the industrialization of nature, begins with ignoring seasons.

Agribusiness leads the way. By favoring, for example, specially selected varieties of long-keeping apples, big growers can sell apples all year. By dominating the market with these varieties, over time—out of familiarity and habit—such apples become what we "want." The varieties that do not tolerate warehousing cannot compete because they do not assure profitable income all year, even though the profit they bring per apple may be higher during their season. So traditional fruit and vegetable varieties are bulldozed in favor of the genetic uniformity on which mass marketing thrives.

Buying Fresh Produce in Season

How can you and I halt those bulldozers, or at least slow their progress? By purchasing foods when they are in season. If we show retailers that we honor nature's cycle, farmers will get the message. Remember, it is most profitable for producers and retailers to supply what we want, and if that turns out to be food in season, they will follow the trend. It will then *pay* to plant and sell the higher-quality, diverse fruits and vegetables that do not keep well. The mass-market varieties will no doubt remain, but the gene pool will be strengthened by the addition of commercially viable seasonals.

The benefits to us as consumers are threefold. First, produce in season is less expensive: commodity prices are held in check by the forces of supply and the desire of producers to avoid the bins and chutes of the low-price processing market. Producers' costs for storage and retailers' costs for long-distance transportation are likewise minimized. The energy and environmental costs of conveying produce thousands of miles are other consequential concerns. According to one researcher, it takes an energy expenditure of 435 calories to fly a five-calorie strawberry from California to New York.[25]

Second, peak of season means peak of taste, whether the produce is corn, melon, or new potato. A Gravenstein apple—available only a few months of the year—awakens the taste buds; a mass-market apple, stored and then sold out of season, devalues flavor. While better taste is not strictly speaking an argument for biodiversity, the inevitably superior flavor of truly fresh produce is one of its strongest appeals. If we hold out for what really tastes good, rather than let our senses be lulled to sleep by advertising and the ready availability of lower-quality goods, we help save endangered types of food plants.

Third, food in season is safer, because it has likely had a

What's In Season?

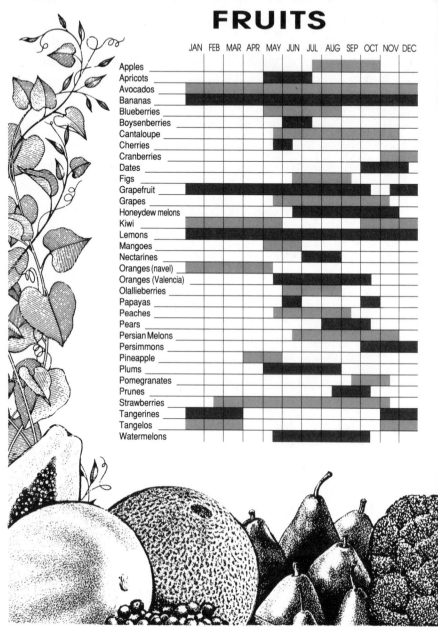

FRUITS

	JAN	FEB	MAR	APR	MAY	JUN	JUL	AUG	SEP	OCT	NOV	DEC
Apples												
Apricots												
Avocados												
Bananas												
Blueberries												
Boysenberries												
Cantaloupe												
Cherries												
Cranberries												
Dates												
Figs												
Grapefruit												
Grapes												
Honeydew melons												
Kiwi												
Lemons												
Mangoes												
Nectarines												
Oranges (navel)												
Oranges (Valencia)												
Olallieberries												
Papayas												
Peaches												
Pears												
Persian Melons												
Persimmons												
Pineapple												
Plums												
Pomegranates												
Prunes												
Strawberries												
Tangerines												
Tangelos												
Watermelons												

Source: California Fresh Produce Guide, Publication No. 28 (Los Angeles: California Department of Food and Agriculture.)

VEGETABLES

	JAN	FEB	MAR	APR	MAY	JUN	JUL	AUG	SEP	OCT	NOV	DEC
Artichokes												
Asparagus												
Beans (dry)												
Beans (snap)												
Beets												
Broccoli												
Brussels sprouts												
Cabbage												
Carrots												
Cauliflower												
Celery												
Corn												
Cucumbers												
Eggplant												
Endive												
Escarole												
Garlic												
Kohlrabi												
Leeks												
Lettuce												
Mushrooms												
Okra												
Onions												
Parsnips												
Peas												
Peppers (green)												
Peppers (chili)												
Potatoes												
Rhubarb												
Rutabaga												
Spinach												
Sweet potatoes												
Squash												
Tomatoes												
Turnips												
Watercress												

lighter dose of farm chemicals. Much out-of-season produce is imported, potentially from countries where use of pesticides and herbicides is poorly regulated. Consider the mid-winter tomato: up to 50 percent of the tomatoes retailed in the United States from December to May come from the Culiacán Valley in Mexico, where crops (and workers) are sprayed by airplane every four to seven days. Besides being sprayed with pesticides and herbicides, produce is often subjected to chemicals that alter natural processes for marketing purposes. Alar is the most obvious example. Until recently, this toxic chemical was widely used in the apple industry to trick acres of fruit into ripening simultaneously and coloring into likenesses of the "Snow White" apples of magazine advertisements.

While some in industry argue that chemicals such as Alar lower the cost of food (by trimming labor time), they leave out of their calculations the health-care costs of those who are exposed while cultivating, picking, and processing the crops, or while consuming them. Nor do they consider the unknown effects on human health of lifelong consumption of produce that has been stored in gas-filled warehouses or irradiated so that it can be sold weeks or months past its prime. These are the hidden costs of playing with nature's clock.

Eating by season is easy to assimilate intellectually. But what do you do when your child brightens at the sight of fresh peaches in February—and your mouth begins to water, too?

To inspire myself, and my children, with a respect for nature's patterns, I have developed an analogy. For most people, regardless of religious belief, the end of December is a special

Winter Salad Selections

INSTEAD OF	CHOOSE
lettuce	red and/or green cabbage
	Napa cabbage
	spinach leaves
	escarole
	sprouts (all kinds)
sliced tomatoes	dried tomatoes
or cucumbers	fresh mushrooms
	blanched beets
	blanched celery root
	slivered carrots
	slivered jicama
	slivered parsnips
	marinated artichoke hearts
	marinated garbanzo beans
	apple or orange chunks
	sliced avocado
	pitted olives

time: decorations adorn stores, streets, and houses; aromas of fir trees, wood fires, and spiced cider waft through the air; special music, books, and clothes are brought out of closets. Would the holiday be as wonderful if we kept the Christmas tree up all year, opened presents every morning, or had turkey and stuffing for dinner each night? Hardly. Anticipation and memories are part of the magic of this time of year. Christmas every day just wouldn't be as much fun.

Similarly, the first peach of summer would lose its charm if we ate peaches all year. Nature's foods are meant to be celebrated in season: the rites of asparagus and rhubarb in the spring, cherries and sweet peas in the early summer, corn and melons later on, squash and pomegranates in the fall, persimmons and Brussels sprouts in the winter. There is a lovely rhythm to the seasons of our food—if only we could hear it!

Eating in sync with the seasons also tends to deindustrialize our view of nature by exposing us to the waxing and waning of complex natural cycles. My children have learned about the life cycle of apples, about how the fruit they dearly love begins as sweet-smelling, pretty flowers that grow into bitter, hard, green pebbles that gradually, over the summer, transform into large, sweet, green fruits. Now they are well attuned to the yearly maturation process; the first ripe apple is a momentous event, like Christmas.

Beyond Fresh Produce

We can wean ourselves, and our children, of ingrained habits such as tomatoes in January and grapes in March. The key is to pull in gradually, almost imperceptibly, on our seasonal boundaries. Moving slowly staves off revolution within the family and provides time to explore new culinary horizons. If you crave tomatoes in midwinter, fetch some canned sauce off the shelf or your fresh-frozen sauce from the freezer, then boil some pasta. Or marinate dried tomatoes in olive oil and toss them in a salad. Grapes, of course, dry into raisins, an easy lunchbox alternative. For dessert, try gently stewing them with other dried fruit, cinnamon, and cloves. Vary your routines: in the winter, make coleslaw instead of a tossed green salad; hot cereal with dates instead of cold cereal with berries. Let your imagination, instead of agribusiness marketing strategies, guide your diet.

Yes, it is better to turn your cart toward the frozen and canned-food sections of the supermarket in midwinter than to buy fresh produce flown in from other countries. Behind imported fresh produce lurks the potential for great damage to the planet's genetic resources. For example, I was recently in Chile, a country that is now a major source of fruits and vegetables for the United States in the winter because when it is winter here, it is summer there. Strolling through a Chilean farm market with friends who are not aware of biodiversity issues, I learned that in only a few years many traditional fruits and vegetables have become almost impossible to find. They have been crowded out by produce destined for the high-paying American market. When we demand typically American produce out of season, we create a biological crisis elsewhere, putting traditional foreign crops under threat of extinction.

And we do have good alternatives, in the form of frozen, dried, and canned foods. Nutritionally, the differences between these food products and the fresh items are not great—

if we exert a little common sense in our selection and cooking. For example, do not let frozen foods thaw and drip away their nutrients; make use of the water packed with canned vegetables; and do not buy heavily sugared fruits. Of course, some of these foods may have unwanted chemicals, so read the labels carefully to avoid unnecessary additives and preservatives.

Put simply, the idea is to buy what is local and fresh, or select from among the preserved alternatives. This strategy will not dramatically change the world, but it puts an additional subtle pressure on farmers and retailers to adopt practices more in harmony with nature's cycle. It is not easy for an industrialized people such as ourselves to opt to do without, or to submit to natural forces rather than try to control them. But if we compromise a bit, we can reduce our negative impact on the natural world. We have everything to gain, and we have already lost a lot.

Expanding the Market

Retail marketing in the United States is a subtle and complex art whose form today is shaped by the seemingly conflicting influences of consolidation and specialization. Consolidation bodes only ill for biodiversity; specialization, however, can be a useful tool in the planet's protection.

Consolidation is familiar in the form of the corporate merger, that process by which large companies grow ever larger by swallowing up smaller ones. Supermarkets are part of the trend toward consolidation; both individual stores and store chains have expanded in recent decades. Megasize chains are the larger evil, biologically speaking, because they tend to standardize what they buy to simplify management tasks, achieve economies of scale, and leverage the best deals with suppliers. Such standardization codifies what is purchased to large quantities of a few items from big farms. That means less diversity in what is planted and, eventually, less diversity for our planet.

Specialization is the other powerful retailing trend of the 1990s. It, too, is compelled by concentration of capital—in this case, in the hands of consumers. Groups of consumers with similar tastes constitute a target market for retailers who sell goods that might satisfy those tastes. People naturally gravitate toward smaller shops that seem to carry specifically and exactly what they want to purchase. Our local shopping mall, for example, boasts one clothing store for petite women, another for pregnant women, and a third for older women. Department stores and supermarkets try to counter the specialization trend by establishing boutique sections that mimic the specialty stores. One department store in our mall divides its stock of women's clothes between two floors and eight departments. A nearby supermarket underwent renovation to install a bakery and a delicatessen and to remap the produce section to achieve a softly lit, countrified atmosphere.

Retailers who follow the specialization trend are successful if they respond effectively to the most compelling need of a specific sector of the market. What is our most compelling need in selecting a grocery store? According to a recent supermarket industry report, surveys show that consumers in *every* demographic group rate the quality of fruits and vegetables their number-one criterion in choosing a food store.

The subdued lighting and soothing music piped into supermarkets cannot, however, hide the tastelessness of industrial-strength fruits and vegetables. Some supermarkets offer fruits and vegetables outside the mainstream of agribusiness provisions, but with certain crops, or within many supermarket chains, no significant choices exist. Supermarkets simply do not specialize in produce.

So if we are in search of quality fruits and vegetables, where should we go? A number of excellent alternatives exist, including urban micromarkets, farmers' markets, roadside stands, and health-food stores. Each is a resource for purchasing tastier, nonagribusiness produce that is—not incidentally—apt to help retain global biodiversity. While not all of these alternatives may be available in your neighborhood, few areas of the United States are wholly dependent on supermarkets.

Urban Micromarkets

I once asked four New Yorkers I met at a conference if they had ever seen red or yellow bell peppers, and if so, if they knew where to get them. Independently, all four replied, "Korean stores."

Scattered throughout New York, as is true of most cities, are small, multipurpose shops run by ethnic entrepreneurs: Korean, Chinese, Italian, Hispanic, East Indian—the types are as diverse as the peoples of the world. These stores offer city dwellers the conveniences of nearby location, long hours, and minimal check-out lines. More importantly, they provide the highest-quality foods of their respective cultures: baked goods, spices, sauces, dried beans, pastas, and grains unmentioned in *Joy of Cooking*. Many also manage to secure specialty produce from small, independent growers: Japanese egg-plant, bitter-melon squash, burdock root, fresh horseradish, tomatillos, tamarind, jicama, and other items essential to foreign cuisine.

In most cases, the shop owners—rarely raised on super-market produce themselves—have high expectations of what constitutes acceptable food. Therefore, their produce is often of higher quality than what is found in supermarkets. As with other independent sources of food, these shops contribute to maintenance of the edible-plant gene pool. They provide a vi-tal link between those of us who wish to eat fresher, more var-ied, and less poisoned food and those who grow such foods and need a market.

By shopping at urban micromarkets we increase the pres-sure against monoculture. At the same time, we expand the market for "exotic" foodstuffs. Eventually, of course, super-market managers discover that we are opening our wallets down the street to buy radicchio or lemon grass or Padilla chilis, and they add these items to their shelves. This is good for the suppliers of these foodstuffs and good for the gene pool, although it is hard on the micromarket owners—be-cause their market niche is forever under assault.

One seeming drawback to shopping at a micromarket is the higher prices often charged for basic commodities. Part of what we pay for in these little shops is their convenience. More importantly, however, we are paying for a better selection of superior food. After all, are pesticide-laden, cardboard to-matoes really a bargain?

Infiltrate the Neighborhood

If faced with the conflict of yearning for produce or other foods from a store that is just a few too many miles away, try enlisting a friend, relative, or neighbor in a little "swap shopping."

Temptation is the key to securing his or her interest. Take your potential shopper a gift of quality produce or exotic foodstuffs from your favorite store. Follow this with more gifts, to raise the person's level of consciousness and expectation of what constitutes good food, until he or she inquires about your source. Then casually lay out a plan to swap shopping expeditions. The plan might be as simple as each of you making a trip to the chosen store once a month. You can build from there as time and interest allow. Through your initiative, not only will you both eat better, but you will have doubled your contribution to preservation of biodiversity.

Farmers' Markets

Less convenient but less expensive than micromarkets are the farmers' markets found in cities large and small across the country. In terms of hours of accessibility, some markets, such as Baltimore's Lexington Market, operate daily. Those in smaller cities or towns are typically open just one or two days a week, sometimes for just a few hours. Some of these also close down during the winter.

The temporal inconveniences of farmers' markets are somewhat ameliorated by their hallmark feature: the person who labors to produce the food places it directly in the hands of the consumer.

Thus, farmers' markets recreate opportunities for the bygone interactions of the marketplace and the pleasure of purchasing from the source. Here, earth-stained hands bundle beets and make change, and farmers discuss with pride the varieties of produce on display: which peaches for eating, which for canning; the uses of different peppers. Gone are the middlemen—brokers, distributors, wholesalers, retailers—of the modern supermarket industry. Absent, too, are the packaging experts, advertisers, management executives, and stockholders, who usually take a piece of the action. Because farmers' market goods go directly from soil to cook pot, the prices are usually low, reflecting only the provider's actual costs and the market demand for the provisions.

While the food is good and the prices are right, farmers' markets do inflict other costs on consumers: because there is little or no intervention from intermediaries, nor attempts to manipulate nature's cycle, the items available vary from day to day, and even from hour to hour. Living by the harvest cycles in this way certainly reconnects urbanites with the natural world. However, if you are planning a dinner party whose success depends on procuring four large eggplants, twelve medium-size artichokes, and two pints each of strawberries, boysenberries, and blueberries, and farmers' market might be your undoing.

On the other hand, building a menu based on current crops is how most animals—indeed, most people on this planet—live from day to day. With a good cookbook or advice from the person who sold you the kohlrabi or blue potatoes, you should be able to prepare a tasty repast without much trouble. For example, why not banana squash instead of eggplant, stuffed mushrooms instead of artichokes, and a kiwi tart instead of mixed berries?

Roadside Stands

In communities where remnants of farmland still ring an urban area, food providers simply sell the products of their labor by the side of the road. When traffic justifies, these stands can resemble small farmers' markets, acting as outlets for several local farmers—to the mutual benefit of growers and consumers.

At an even more complex level of development, roadside stands can become a kind of country cousin to the urban micromarket. There are several of this variety near my home. The one that draws most of my business sells locally grown produce along with goods purchased through wholesalers, thus blunting the sharp cycles in local availability of some foods. The market owner further broadens his establishment's appeal by augmenting the produce with staples such as milk, coffee, and ice cream, as well as indigenous items such as locally produced eggs, jams, spice mixes, and fresh and dried flowers. But the focus is still on produce—the best of whatever can be had. Unlike the supermarket buyer, the roadside-stand owner is not locked into long-term contracts with suppliers; he or she is free to try small quantities of odd items to see what sells. This enables local growers to expand the gene pool a bit by serving markets for nonstandard produce. And since local people can both buy and sell at these establishments, the community benefits on both sides of the exchange.

Health-food Stores

Without debating the meaning or utility of the term "health food," good food—and organic produce—is readily available in many of these stores. Beware, however, the stores that emphasize food supplements rather than fresh goods; in these establishments, produce offerings can resemble supermarket fare, and low turnover can betray freshness. But in states where the term "organic" is backed by certification and inspection, wonderfully poison-free produce makes it worth the trip to the health-food store. And, since it is the older, traditional varieties of fruits and vegetables that succeed in organic (chemical-free) growing environments, while modern varieties—bred to depend on herbicidal crutches—do not, patronizing these stores is a simple, good move in defense of biological diversity.

The shelf goods in these stores also support biodiversity. Like ethnic markets, these establishments are apt to stock several kinds of rice, many other grains, numerous types and grinds of flours, wide varieties of dried beans, and a spectrum of herbs and spices. Many of these items are outside the norm of American cuisine, coming as they do from other cultures. Though we may know intellectually that most of the rest of the world basically eats grains and beans, the supplies in these stores vividly convey the breadth and potential of such a diet. Universes of flavor await those who dabble, however timidly, in these ancient and neglected staples and spices. Each time we venture beyond supermarket kidney beans to try adzuki beans, buckwheat groats, orange lentils, or amaranth cereal, we help preserve for the future a more diverse set of protein-providing plant genes.

All four alternatives to supermarkets provide opportunities for expanding our culinary palates while contributing directly to the conservation of biological diversity. It is unlikely, however, that any one of them can match the overall convenience of a large supermarket, so using them may pose logistical problems. We may have to shop twice: once for supermarket staples and once for specialty items. For some busy people, the extra shopping or extra stop is a critical barrier. For me, the thought of having to unstrap and restrap two fidgeting children from their car seats is oftentimes enough to restrain me from stopping for local tomatoes while on my way to the supermarket.

While there is no perfect solution, two ideas may help you to broaden your shopping range without experiencing major inconvenience.

The first idea works in two-adult households. It involves dividing the shopping list so that each person is responsible for a different type of store. The shopping list on my refrigerator, for example, has a section for the supermarket and a section for the local roadside market. At the appropriate time, the list is torn in two: half goes into my hand, the other into my wife's.

The second idea is predicated on changing your shopping schedule. If your local roadside stand, urban micromarket, or health-food store also carries staples, it may be possible to reduce supermarket excursions by half—making the trip once every two weeks instead of once a week, for example. The few-cents increase for milk or coffee is offset by the eliminated trip. A corollary strategy involves shopping at different stores on alternating weeks—at the supermarket one week, then an alternative store the next. A bit of rearranging of storage shelves and the refrigerator at home should permit longer intervals between shopping trips. Just be sure to store the items properly and remember to check the expiration date on packaged foods. Many foods, from rice to eggs and from cereals to jam, keep well for weeks and months. Laundry soap, toilet paper, and garbage bags are items that never go out of stock and never spoil—why let them compel you to the supermarket each week?

Both of these shopping strategies require more planning than the simple routine of gliding a cart around the familiar aisles of the same supermarket each week. But the true price of the more simplistic approach is inferior food for us now and a dangerously depleted living-gene bank for our future. "Food store" does not have to mean "supermarket." As we broaden our shopping territory, we help preserve the world's genetic heritage.

GROWING YOUR OWN

In a recent Harris poll, Americans reported gardening as their favorite form of recreation. Another recent survey showed that in the summertime, a third of the fresh tomatoes eaten in the United States are plucked from home gardens—proof that we are still not won over to cardboard supermarket tomatoes. For those of you already practicing America's favorite pastime, this chapter integrates the issue of biological diversity with what you plant in your backyard plot. For those of you who look over the fence with wonder and envy at your neighbor's red-ripe tomatoes, it offers words of encouragement to try growing your own.

As veteran gardeners know, what to plant is the number-one consideration in getting going each year. For most Americans, "seed source" is synonymous with "supermarket" or

"Burpee." And that is a problem for the planet because these sources are controlled by big corporations. According to the Seed Savers Exchange of Iowa, more than sixty North American seed companies were recently taken over, and more than fifty others went out of business between 1984 and 1987.[26]

Corporate Control of Seeds

What is wrong with big corporations owning seed companies? Don't they just achieve greater economy and wider distribution, especially of new and "better" varieties? Indeed, they do, but this is not something to celebrate. Corporate control of seed production and distribution has wide-ranging repercussions.

DIMINISHED DIVERSITY

Big corporations tend to offer a limited number of seed varieties. In any kind of business, big corporations always do what they can to reduce or eliminate competition. When the competition is another company's seeds—actually precious packages of genes—they end up with a powerful interest in reducing the diversity of the gene pool, all for the sake of market share.

PIGGYBACK SALES

Through the well-established marketing channels of seed selling, large corporations can easily piggyback sales of other, more profitable farm products. The other products, however, tend to be noxious pesticides, herbicides, nematocides, and fertilizers. Not incidentally, the seeds corporations push are hybridized—and therefore dependent on agrichemicals. In addition, most hybrids do not occur in nature and will not reproduce "true" from seed. Unlike open-pollinated varieties, there is no purpose in letting plants go to seed to collect them for planting the next year. Instead, hybrids automatically create return customers for the corporations that sell them. Between assured sales of agrichemicals and seeds, there is a large profit motive to promotion of hybrids.

LACK OF FLEXIBILITY AND LOCAL ADAPTATION

Much to the frustration of many a gardener, the most readily available seeds, the hybrids, are poorly suited to local growing conditions. The less readily available, open-pollinated plants make use of their natural genetic plasticity to adapt to the unique growing conditions of your backyard. Seeds saved from these plants produce new, similar plants the next season. If you save seeds from your best plants each year, you will soon have varieties ideally suited to your yard, your microclimate, and your gardening habits. Your neighbors' success with plants they have been growing for years may be due to just this kind of selective breeding. Promotion of genetically uniform, chemical-craving hybrids inevitably results in the extinction of open-pollinated varieties. And nothing is as effective as extinction in narrowing our selection of what to plant.

INSENSITIVITY TO LOCAL NEEDS

Of the many ways we interact with plants, one act is crucial to our planet's genetic legacy. Faced with a patch of carrots, corn, wheat, or whatever going to seed, we develop a mental image of the next crop, then we carefully select seed from the plants closest to the ideal: we shape the world around us with our minds. The danger with the corporate seed-selection process is that an enormous distance separates the person generating the image from the person ultimately planting the seeds. The idea of the desired plant is not created in the mind of the Filipino rice farmer or the Ghanaian peanut grower, both of whose images would produce plants suited to the special requirements of their areas or people. Instead, the image is captured in memos from Basel or sales presentations in Minneapolis. The seeds promoted by big corporations relate to the priorities of commercial rather than home gardeners.

Through corporate development and disbursement of hybrids, agribusiness robs us of innumerable seed varieties, limits our choices, and precludes a rare and precious chance for farmers to interact with nature in a positive, satisfying, and mutually beneficial way.

Seed Saving

Yet it is not too late to help reclaim our planet's genetic legacy. Through seed saving, the question of "seed source" can be answered in your own backyard. The tradition of passing down heirloom seeds can be carried on by any of us, at any time, no matter where we live—even in New York City, where a friend of mine grows hot chili peppers in a window box on his fire escape, the seeds passed down to him by his family in Louisiana.

There are just five steps (one of them optional) to joining this wonderful tradition that ensures our current favorite pastime will thrive in the future.

1. Plan and plant a garden. The key to planning a garden, if you do not already have one, is to start with the absolutely smallest plot you can imagine. We amateurs inevitably overextend ourselves, fail, and quit. I have grown into a garden five feet wide and twenty feet long. Even at this modest scale, I never quite get the whole thing planted. I always tell people I am letting the "south forty" go fallow. In my case, the forty is in inches, not acres.

Buy an excellent gardening book; there are many around that are worth their price. Expect to ignore 50 to 90 percent of what the book says. (Your plants won't know, and at least you'll have a reliable resource.) The book I have is John Jeavons's *How to Grow More Vegetables* (Berkeley: Ten Speed Press, 1979).

2. Obtain seeds. Avoid the ubiquitous spin rack at your supermarket and the plastic trays of seedlings at your hardware store that will look so tempting on a nice spring Saturday. Instead, at least for some of your planting, buy open-pollinated seeds. These are generally available from smaller seed companies, mostly by mail order, but one store not too far from my

Saving Seeds for Your Future

"Always look at your plants with seed saving in mind. Don't just look at the fruit; look at the whole plant. Select several plants to save seed from, not just the best looking or largest one. This will give your seed greater genetic diversity and is key to your plant's continued ability to adapt to a variety of conditions. Remember, your vegetable plants and their seeds change slowly in response to environmental and genetic factors. Select characteristics for the future: size, flavor, earliness, ability to survive a short season, disease resistance, drought resistance, insect resistance, trueness to type, color, shape, lateness to bolt, hardiness, and storability all can be selected for.

"Take precautions when drying and storing your seed, to preserve the seed's vigor. Vigor is your seed's ability to germinate rapidly with good disease resistance and is readily destroyed by

house now sells seedlings started from nonhybrid seeds; so ask around your community.

Explore other people's gardens, too. Find out what is behind Uncle Charlie's fabulously tasty tomatoes. It may be a European conglomerate, but just maybe he saved seeds from Grandmother's patch. Your best bet is with nearby relatives and neighbors, since their seeds are probably locally adapted and thus more likely to survive in your garden. Experience has high temperature and high moisture during storage. Seeds can be dried on a screen or on wax paper in the sun, by sealing them in an airtight container with silica gel, by putting them in a pilot-lit oven with the door ajar, or by using a food dehydrator. In the oven or dehydrator, the temperature must be controlled so that it does not exceed 95° F.

"Your seeds must be completely dry (less than 8 percent moisture) before storage; in other words, they should break instead of bend. Dried seed should be stored in a completely airtight container at as low and constant a temperature as possible. Specifically, put them in small quantities in individual envelopes; write on the outside of each envelope information about the plant variety. Put the envelopes into any glass jar that has a rubber-gasketed lid. Screw the lid down hard to make the container airtight and moisture proof, and store the jar in your freezer. Then look forward to next year." *The 1988 Edition of the Seed Savers Exchange* (Decorah, Iowa: Seed Savers Exchange, 1988), p. III.

taught me this lesson. I planted seeds of my friend's Louisiana–New York chili peppers, but each plant died as soon as it grew far enough above the soil to realize it was in an entirely different zip code. On the other hand, when I planted amaranth seeds from a plant a friend had grown a few miles away, the plants grew big and strong—despite the fact that I had no idea what they should look like or how to care for them. They thrived because they were at home.

3. Involve your family. This is the optional step. Not everyone has children or wants them messing around in their garden. And many people use gardening as a time for solitude, quiet, and reflection—the antithesis of time with children and a precious commodity in our loud, hurried world.

But gardening provides an opportunity to instill in the future custodians of our troubled planet a feeling of connection with the earth and vegetation. This is at least as important a value as remembering to use the recycle bins or to share toys. Perhaps more important.

Aside from a chance to love soil and plants, gardening can be an important antidote to TV time. Too often children expect action and resolution within twenty-five minutes, the span of the average television program. When I gave my three-year-old daughter bean seeds to plant, she poked a few into the earth and then stood back, hands at her sides, solemnly staring at the depressions her fingers had made. After a few minutes, when I asked her what she was doing, she said she was waiting for her beans because she was hungry. Gardening can help bring our sense of time, and our children's sense of time, back into better harmony with nature.

One strategy for gardening with your children is to give them a special plot of their own, or even just a planter box. This gives them a sense of ownership, responsibility, and pride; permits you to have neat rows elsewhere; and reduces your potential for heartbreak when they thin out every single tomato seedling, as happened to me recently.

4. Save your seeds. Resist the temptation to rip out all of your plants at harvest's end. Instead, observe your plants while they are producing and identify those exhibiting desirable characteristics. These are the plants with "keeper" seeds, so let them go to seed. Some plants, such as tomatoes, make you labor for their seeds; some, such as squash, practically drop them in your lap. Do not be afraid to try out a seed-saving project. Most seeds are pretty hardy. Were they not, the world would be barren by now. At worst, you will have to buy some seeds again next year; at best, you will gain a perpetual, custom-tailored seed set that you can someday pass along to your grandchildren.

5. Share your seeds. After a few seasons, when you have found a plant you like and that seems to like you—one that has developed into something you are proud of—start your own Johnny Appleseed program. Make up little seed packets and distribute them to family, friends, neighbors, and coworkers. While some folk will not understand what you are up to, you will invariably run into at least one enthusiastic gardener who will reward you with special admiration and appreciation for your heritage-seed-saving activities.

Even a tiny home garden can provide myriad benefits: recreation, healthier and tastier food, and a closer connection with the rhythms of the earth. The most important benefit, however, is an immediate and effective contribution to the preservation of our planet's threatened genetic heritage.

Seed Savers Exchange

The Seed Savers Exchange (SSE) is an organization of gardeners who locate and maintain historic varieties of vegetables and fruits.

For as long as emigrants have come to the United States and Canada, the gardeners among them have carried along their favorite seeds. Many heirloom plants are still being grown in rural areas and ethnic enclaves of the United States, but given the current economic pressures against small-time farming, elderly gardeners often can find no one interested in growing their unique varieties. These outstanding varieties will soon become extinct unless dedicated new gardeners are found to continue to plant the seeds. Since SSE was founded in 1975, its members have located thousands of old-time vegetable varieties and have distributed more than 300,000 samples of garden seeds. These seeds are typically for varieties that are not available through catalogues and that are, in many cases, on the verge of extinction.

The *Garden Seed Inventory*, produced by SSE as a preservation tool, is an inventory of 215 mail-order seed catalogues from the United States and Canada. It describes 5291 standard (open-pollinated) vegetable varieties and lists the companies offering each one. Softcover ($17.50) and hardcover ($25.00) editions are available postpaid from the Seed Savers Exchange, Rural Route 3, Box 239, Decorah, IA 52101. The exchange also offers, free for the asking, a four-page brochure on seed saving.

FAST FOOD: PERVASIVE
FOE OF DIVERSITY

To discuss food without mentioning fast food would be almost un-American. Fast food is nothing if not pervasive in the United States, where 50 percent of us live within fifteen minutes of some chain outlet. And it is becoming familiar across the globe: in 1985, a new McDonald's opened its doors somewhere on the planet every fifteen hours.[27]

Fast food is a wonderful and a terrible thing. On a typical day, nearly half of all adult Americans patronize some type of food-service establishment, reports the National Restaurant Association. And according to industry association projections, roughly 80 percent of all dollars spent on eating out in 1990 will pass over the counters of fast-food establishments.[28] While we love the convenience of fast food, we also rely on its comforting predictability—we want, expect, and, indeed, receive the same shape, flavor, and texture of burger every time.

But the burgeoning popularity of "limited menu" restaurants (as they are known in the industry) results in a declining need for diversity in produce and other foods as ingredients. The short grocery list of this $70 billion-per-year industry[29] exerts tremendous pressure on what is planted and harvested in the United States, mostly with the effect of reducing total biological diversity.

Like many modern inventions, fast food has several immediate benefits. First, it is fast. Traffic jams, multiple demands of work and child care, and twentieth-century impatience nurture our interest in instantaneous meals. Second, it is portable. We can eat in the car, on the job, or at the dining-room table. Third, it is consistent. No surprises. Kentucky Fried tastes the same in New York or DeKalb, Peking or Panama City. Fourth, it is fabulous for kids. Standard restaurants are stages set for embarrassing experiences with children. Overturned chairs, spilled food, finicky palates, whines, and tantrums—the opportunities for humiliation drive even the hungriest of parents past establishments that offer scallops brochette in search of burgers and a relatively peaceful meal.

The trade-offs for all this convenience are serious. On the personal level, fast food is devastating to good nutrition. High in fat, salt, and total calories—but low in a range of essential nutrients—too much is a health hazard. *The Fast-Food Guide* (New York: Workman Publishing, 1991), by Michael F. Jacobson and Sarah Fritschner, is a great resource for information on the ingredients and nutrients in the foods of the top fifteen U.S. chains.

On the global level, fast food is injurious to genetic re-sources. Fast-food restaurants require huge quantities of cer-tain food items at a constant rate of production and the lowest possible price. Given the parameters of uniformity and pre-dictability within which they must operate, the industry's al-liance with agribusiness seems inevitable, almost natural. Hand in hand, these two giant industries strive to limit incon-sistency. But their combined interest in producing and secur-ing flawless, uniform, long-lasting vegetables and fruits ulti-mately narrows the plant gene pool. The competitive nature of the business virtually ensures that each burger's slice of to-mato was born of the toughest, cheapest, least flavorful, and, consequently, least genetically diverse tomato variety avail-able.

At least the tomatoes for American burgers are—for the most part—grown on American soil. But as our appetite for fast food increases, and as this American industry spreads abroad, we create genetic havoc elsewhere, driving farmers in other parts of the world to abandon traditional crops in ser-vice to an imported, highly specialized menu.

The fast-food industry's clamor for cheap beef produces a particularly awesome ecological aftermath in the tropical rain forests of the world. Every year, in the Amazon Basin and else-where, an area of tropical rain forest equal to the size of Penn-sylvania is cleared for grazing fast-food-chain cattle—despite the fact that the land is so poorly suited to this purpose that it must be abandoned after just a few years of use. And despite the fact that more than half of all plant and animal species on

Growing Up on Fast Food

Fast-food restaurants are designed as havens for children. Tables and sometimes chairs are bolted down, carpets are absent, and no one says "shush" when everyone's squirming and yelling. As for the limited-choice menu, children are notoriously (heartbreakingly!) unadventurous when it comes to eating. They learn to love the familiarity of fast-food menus. Once, while on a beach on Kauaì, I asked a child in drooping diapers what he was enjoying most about his family's trip to Hawaii. He said it was that they got to go "berking" every day. Imagining some new water sport, I asked the boy's brother when the family would be "berking" again, and if I could watch. He gave me a puzzled look and said he supposed they would be having dinner as usual at Burger King that night, but he would have to ask his mother if I could watch.

Children's allegiance to the familiar poses a subtle, insidious problem. The more frequently they eat fast food, the more they come to expect and demand highly processed, quickly cooked, nutrient-poor food, and to reject new foods, flavors, and styles of presentation. I am fearful that, if we are not mindful, the omnipresence of fast-food restaurants may completely industrialize their relationship with the substance of life. Food-getting for

earth live in rain forests. To put this in perspective, the Amazon River system, for example, hosts about eight times as many living species as does the Mississippi River system.[30] Those clearing the rain forests for profit today overlook the fact that these primary forests may not be able to regenerate them is not plunging hands into warm, crumbly soil, but shouting out of a car window to a plastic signboard.

Yet our children will someday have to deal with the complex issues of food, agribusiness, economics, and politics—how to feed the planet's population in the decades to come. We owe it to the planet to prepare them for the task.

Become an agent for change within your family by reducing or even eliminating fast-food propaganda from your house. If you are not on guard, the industry will invade your home and colonize your children's minds with images of clowns and characters designed to hook them on chain food. Ban giveaways and geegaws that give the industry an inroad to your children's ability to make sound dietary choices.

The next time the subject of eating out comes up and your children start clamoring for a meal from the chain that has been singing its name to them on television all day, be ready to propose an alternative. Be creative in exploring various options—from different kinds of restaurants to picnics around the fireplace or even pizza for breakfast—to match their desire for novelty. It is not easy to combat millions of dollars' worth of persuasive advertising, but your children's health and our world's genetic heritage are worth the effort.

themselves after such an onslaught. As we destroy these habitats, we destroy untold numbers of species whose interaction with the earth's biosystem is not yet known or understood.

The Burger King boycott of 1987 was effective in compelling that company to stop buying beef from Latin America, beef that was cheap in price but expensive in ecological consequences. The boycott's success illustrates again that we consumers can influence the food industry by minding how we spend our dollars. Too often we are duped by corporate spokespeople who tell us that their company dishes up only what the people demand. In the case of rain-forest beef, the boycott illustrated that we, the people, believe that environmental priorities supersede price.

Participating in a boycott of one chain does not work the fast-food addiction out of our system, however. What can we do in an environment that bombards us daily with inducements to consume, immediately, billboard-look-alike meals? We can try out alternatives. Gradually. Just as with dieting, a crash program never works. But there are ways to make subtle, incremental changes that can create a new lifestyle.

One option is to alter your menu. The next time you find yourself at the drive-up window, aim for a more diversified diet. For starters, wean yourself from hamburger. Explore other items on the restaurant's menu: chicken sandwiches, broiled fish sticks, plain English muffins, scrambled eggs, coleslaw, baked potatoes. Your body will thank you, and so will the planet.

A second option is to switch restaurants. Instead of stopping for conventional fast food, order a take-out pizza (plain cheese or with vegetables). Call the Italian restaurant down the street and ask them to prepare an order of lasagne, minestrone, or pasta with tomato sauce. And try the delicatessen for cold salads, sandwiches, and, often, spit-roasted chicken.

Ask the restaurant over the phone if it has your favorites the first time you call, then pick up a copy of the menu to keep. To avoid clean-up at home (after all, lack of dishes is part of fast-food convenience), occasionally use paper plates and napkins.

A third option is to stay home. By that I mean, when you go grocery shopping, pick up provisions for quick meals. Load up on frozen muffins and cups of yogurt for fast, portable breakfasts. For lunches, buy bagels and cream cheese, sliced turkey and crackers, fresh and dried fruit; if you have access to a microwave at work, your possibilities are unlimited. For dinners, stash pizzas, burritos, ravioli, and vegetables in your freezer.

Fast food is here to stay. As the pace of our lives heats up, so does the frequency of our fast-food meals. Perhaps we will not escape McDonald's clutches altogether, but we do not need to give in entirely, either. The more diversity you add to your menu, the more genetic diversity will be sustained on the planet.

STEPPING OVER
THE THRESHOLD

This book suggests a number of simple things we each can do in our daily lives to ensure that the foods we dish onto our plates help maintain rather than destroy the genetic diversity of the earth's food plants and animals. Each of the home-oriented actions recommended, from modifying shopping habits to planting nonhybrid seedlings, will have an impact.

Yet more could be done. This chapter looks beyond the genetics of the foods we bring into our individual kitchens to suggest actions that can help spread word of the biodiversity crisis to our extended families, our communities, and our nation.

Holidays

Holidays are family times and natural occasions for recalling basic values. Many holidays, both secular and religious, draw heavily on nature for symbols and rituals, making them ideal occasions for co-opting as forums for simple lessons about nature.

Thanksgiving is a perfect holiday for this purpose. Historically, Thanksgiving celebrates an early stage in the imperial conquest of this continent. Today that purpose is forgotten as we focus on gathering family together to feast, give thanks, and celebrate good fortune. It is easy to introduce a consciousness of genetic diversity into this holiday, because our ancestors' feast celebrated the abundant natural resources of what was to them a new land. While the origin of the turkey is subject to scholarly debate, the rest of the traditional meal reflects the genetic diversity of the Americas: cranberries are solidly New World; potatoes, sweet potatoes, corn, many kinds of beans, pumpkins, and squash all trace their beginnings to South and Central America.

Each of these foods has a genetic-diversity story behind it. Even the briefest discussion of each food — how it came to be part of the ritual meal, how it symbolizes a biological tie to the past — could inspire deeper respect for the earth's genetic resources.

To go one step further, you could add to your typical Thanksgiving menu new dishes that eloquently bespeak diversity, such as a side dish of quinoa, an ancient American grain now coming back into favor. For an even greater challenge, you might prepare a feast of local products only, or, better yet, the products of your own hands: freshly harvested pumpkin and potatoes from your garden, with green beans, berries, and corn preserved since summer in Mason jars or your freezer.

If each year at Thanksgiving, as we gather at the table with

Kernels of Importance

Margaret Visser wrote in *Much Depends on Dinner* (New York: Grove Press, Inc. 1986, p. 30, 31) that "American Indians in their many different languages, always spoke of corn as 'Our Mother,' 'Our Life,' 'She Who Sustains Us.'

"When Christopher Columbus arrived in America he immediately noted American corn and its immense importance for the Indians. On November 4, 1492, Columbus disembarked on the island now called Cuba. He was met by Indians who hospitably offered gifts, the most sacred and prodigious substances they knew: tobacco and corn. The word they used for the latter, in the Haitian language spoken on the island, was *mais*. In this manner, and on a single day, these two fateful plants were introduced to Europe and the rest of the world. Columbus's journal for November 5, 1492, reads in an early translation: 'There was a great deal of tilled land sowed with a sort of beans and a sort of grain they call "Mahiz," which was well tasted baked or dried, and made into flour.'

"Over and over again the letters and narratives of the early explorers of America express amazement at the cleanliness, the diligence, and efficiency of Indian farming. In Europe at this time (the fifteenth and sixteenth centuries) seed was sown broadcast, that is in great scattered handfuls, irregularly spaced by the chance fall of the grains. Farmers then waited for the weeds to come up with the crop before any attempt was made to sort through the jumble of growing plants and remove the unwanted growth. The Indians did things differently. Round many of the settlements on the eastern seaboard, fields were laid out with chequerboard accuracy, a mound of earth heaped up in every square. There were not draft animals in North America before the Europeans came; all land-clearance and mound-making was done by hand, with the help of axes and sticks, and with hoes whose blades and points were made of wood, sea-shells, antlers, and the shoulder-blades of large animals.

"The women then moved onto the prepared land. They did all the agricultural labour thereafter, up until the harvest, which in some tribes was allotted and stored in quantities decided by the women. At planting time they poked holes into the mounds, and dropped seeds into them: four or six of maize, and, a little while later, a few of beans, a few of squash. Corn, beans, and squash were always eaten together and always planted together: in Iroquois myth they were represented as three inseparable sisters. When the plants emerged from the hill, the corn grew straight and strong, the beans climbed the corn, and the squash plant trailed down the side of the hill and covered the flat land between the mounds. The spreading squash plant helped to keep down weeds. The coastal Indians would plant a fish in every hill as well; to neglect this was sacrilege, and the corn repaid such a sin by refusing to grow properly. The Indians—who can imagine how?— had discovered that maize needs massive doses of fertilizer if it is not completely to drain the soil of nutrients."

our parents and our children, we eat foods that remind us of our venerable biological heritage and give thanks to the living world that sustains us all, we could establish a new tradition of immeasurable value to our planet.

Recipes and Cookbooks

Recipes and cookbooks written with a sensitivity to the seasonal nature of foods could ease the practice of eating in season into the mainstream. Cookbooks offering recipes for unfamiliar foods—such as quinoa, amaranth, and open-pollinated varieties of produce — might inspire curiosity and competence in using ancient, "rediscovered" foods. One such cookbook is *Cooking from the Garden* by Rosilind Creasy (San Francisco: Sierra Club Books, 1988).

When you develop a seasonal menu or create a delicious new dish from an unusual food, share it with a friend. If you are brave, contact your local newspaper food editor and offer to share your recipe with the community. While you are at it, ask the editor about running a story on local foods, or on the benefits of eating produce in season. Tell the editor about the plight of American apples (86 percent of all known varieties have disappeared in the last ninety years) or about the one gene that controls the stringlessness of green beans. "Planting" a newspaper story this way is often surprisingly easy — and its impact can be surprisingly powerful.

If you are truly a creative cook at heart, write a cookbook! For inspiration, contact your local farmers' market. Chances are good that the market's organizers can provide you with advice on local foods and their seasons (and perhaps even produce samples), and they might even lend support to your effort.

We all need help learning how to dehomogenize our diet; simple recipes can bridge the gap between intention and action.

Curricula

Even if you change what you prepare for meals and reduce trips to fast food restaurants, your children might not understand why you have done these things. How can we pass on these ideas to the next generation?

One answer lies in expanding the relevant science and social-science curricula for kindergarten through high school to include instruction on the importance of biological and genetic diversity. While lessons on the earth's diversity of races, flowers, and trees are common, little attention is now paid to the abundant varieties of food plants. Why not have classroom displays at Thanksgiving that focus on heritage foods rather than on Pilgrim and Native American costumes?

You could make a biodiversity poster or display. Perhaps just one, for your child's classroom, or, if you are more energetic, enough for the school or school district. You can control your degree of involvement. But be aware as you embark on this course that schools are understandably wary of crusades. What could be construed as a direct attack on corporations will be no more welcome than debates over mandatory prayer. To avoid controversy, approach teachers and school officials with displays and lesson plans that focus on local biology. Samples of unusual vegetables or demonstrations of seed-saving methods might be the place to start in the lower grades; high-school students might respond to a genetics- and ecology-oriented discussion of plants' abilities to adapt to their environments.

I was once asked to speak to a local grammar-school class about tropical rain forests. To fulfill my assignment, I brought some posters and snapshots of exotic flora, but I managed to steer the discussion to the fields and forests of the local community. And once I got genetic diversity to the backyard level, the fourth-graders lit up with ideas and enthusiasm.

If you are a planner by nature, write some curricula; if imagination is your strong suit, write a children's book. While either of these volunteer tasks presents a lot of work for anyone, what could be more appropriate to saving the earth's gene pool than to educate those who will soon become its custodians?

Laws and Regulations

This is a hazardous field to enter because it is fenced and plowed and cultivated by those who have farmed it for decades: multinational corporations, food-industry magnates, and agribusiness research specialists.

It is typically easiest (and most immediately rewarding) to

work for change in our own lives and in the lives of our families. Sometimes, however, genetic-diversity issues are decided on a scale that naturally draws us into local, state, or national politics: the government calling ketchup a vegetable is a genetic-diversity issue; allowing fast-food franchises in public schools, hospitals, and airports is a genetic-diversity issue; permitting the patenting of life forms (that is, permitting companies to own the genetic code of a living organism and to control all of its future offspring) is a genetic-diversity issue. These are but a few of the most obvious and troubling causes for action on the political front.

Those of us who are inclined should express ourselves on these issues through the channels of our democracy. We must be adamant that our genetic heritage should not become the domain of any organization or political group. Personal letters always impress congressmen. In addition, of course, there are groups—ranging from organizations that fight off legislation proposed by agribusiness to those that support sustainable farming—through which we can express our views. While these collective efforts are crucial and deserving of support, none of these organizations would want you to send away for a pamphlet or mail in a donation as a substitute for the most essential actions of all: the small, incremental, and long-lasting changes you make in how you eat and live.

The living things with which we share the world depend on us, and we on them. Because the plants and animals of our world communicate in a way that is slow, subtle, and often ambiguous (to our frame of reference), we need to learn to hear them, understand them, and join in dialogue with them. The threat to our planet's life from genetic erosion is real and is getting worse every day. If we slow down, learn to listen, and are willing to change ourselves a bit, we can be part of the process that preserves the earth's precious genetic heritage for our children's children and, hopefully, for the many generations beyond.

Resources for Change

The following organizations in the United States can provide you with information, ideas, encouragement, and materials to use in developing a new kind of Thanksgiving, a cookbook, educational curricula, or a response to pending legislation.

THE AMERICAN MINOR BREEDS CONSERVANCY (AMBC) This nonprofit organization promotes and conserves rare breeds of livestock (horses, cattle, sheep, goats, pigs, and poultry). These animals embody the historical preeminence of North American agriculture and represent a genetic diversity necessary to meet the agricultural needs of the future. They were selected and developed by generations of farmers during an era of sustainability and likely carry the very attributes needed for a sustainable future. The AMBC conducts direct conservation projects, such as a rare-breeds semen bank and population rescues, and provides educational programs, assistance to breed associations, and information on rare breeds. Contact: P.O. Box 477, Pittsboro, NC 27312; (919) 542-5704.

CENTER FOR PEOPLE, FOOD, AND THE ENVIRONMENT (CPFE) This nonprofit research, education, and action center was organized to explore the role of indigenous biological and sociocultural resources as the foundation for equitable and sustainable food systems. It recently published *Food from Dryland Gardens: An Ecological, Nutritional, and Social Approach to Small-Scale Household Food Production.* By presenting examples of indigenous methods along with their scientific principles, this book bridges the gap between indigenous and Western horticulture, demonstrates that both are based on experimentation and observation, and shows how to combine the best of both approaches. Contact: 344 South Third Avenue, Tucson, AZ 85701; (602) 624-5379.

FOOD FIRST/INSTITUTE FOR FOOD AND DEVELOPMENT POLICY Food First is a center for research and education for action. Its mission is to stimulate increased public education and citizen participation in solving critical social problems locally, nationally, and globally — with a particular focus on hunger. The institute is best known for its analysis of the causes of world hunger, the objectives and consequences of development aid, and the crisis of traditional models of development in the Third World. Food First produces books, films, curricula, and other educational materials and operates a speakers' bureau. Contact: 145 Ninth Street, San Francisco, CA 94103; (415) 864-8555.

THE HUMANE SOCIETY OF THE UNITED STATES
(HSUS)   The HSUS's Humane Sustainable Agriculture Program focuses on animal agriculture and the treatment of farm animals raised for consumption. The program educates consumers about humane sustainable farming practices and encourages consumers to patronize farmers who avoid intensive animal confinement and routine use of antibiotics and hormones, and who minimize their use of pesticides and synthetic fertilizers that pollute food and groundwater. Contact: 2100 L Street NW, Washington, DC 20037; (202) 452-1100.

NATIVE SEEDS/SEARCH   This nonprofit conservation, research, and education organization works to preserve the traditional crops, and their wild relatives, grown by native Americans in the U.S. Southwest and in northwest Mexico. It maintains a seed bank and encourages indigenous farmers to continue growing and using their native crops. One of its conservation strategies is to promote these seeds to home gardeners in a catalogue that lists more than 200 varieties. The catalogue is available for $1.00. Contact: 2509 North Campbell Avenue #325, Tucson, AZ 85719; (602) 327-9123.

RURAL ADVANCEMENT FOUNDATION INTERNATIONAL–USA (RAFI–USA)   This private, nonprofit organization focuses on the multifaceted issue of loss of genetic diversity in agriculture. RAFI combines research, education, and international advocacy in its program on biological diversity and the impact of new and emerging technologies on farmers and rural communities worldwide. RAFI works mainly at the international level through the United Nations and its network of organizations, and with national governments. Its "Community Seed Bank Kit" is a manual for international development workers and rural organizers working with farmers and gardeners to preserve diversity. "The Seed Map: Dinner on the Third World" is a wall map illustrating centers of genetic diversity and control of plant genetic resources. Other publications address the threat of biotechnology and herbicide-tolerant crops to sustainable agriculture. Contact: P.O. Box 655, Pittsboro, NC 27312; (919) 542-1396.

SEEDS OF CHANGE   Seeds of Change is the first major national grower and supplier of organically grown seeds. The company has grown out commercial quantities of a wide variety of traditional, heirloom, unusual, and nutritious cultivars in an effort to help reintroduce diversity into the food chain. Seeds of Change acts as a consultant to U.S. organic gardeners and farmers and operates a research and education facility and an organic farm. Its seed catalogue is available for $5.00. Contact: 621 Old Santa Fe Trail #10, Santa Fe, NM 87501; (505) 983-8956.

NOTES

1. Edward O. Wilson, "Threats to Biodiversity," *Scientific American,* September 1989, p. 112.
2. Peter H. Raven, *We're Killing Our World: The Global Ecosystem in Crisis* (Illinois: MacArthur Foundation, 1987), p. 8.
3. Paul Ehrlich and Anne Ehrlich, *Extinction* (New York: Random House, 1981), p. 8.
4. Ehrlich and Ehrlich, pp. xi, xii.
5. Henry L. Shands, telephone conversation, October 29, 1990.
6. Committee on Managing Global Genetic Resources, Board on Agriculture, National Research Council, National Academy of Sciences, *Managing Global Genetic Resources: The U.S. National Plant Germplasm System* (Washington, D.C.: National Academy Press, 1991), p. 9.
7. Rural Advancement Fund International, "RAFI Communique," July 1987.
8. Gary Paul Nabhan, *Enduring Seeds* (San Francisco: North Point Press, 1989), p. 97.

9. Jack Doyle, *Altered Harvest* (New York: Viking Press, 1985), p. 200.

10. Doyle, p. 264.

11. Doyle, p. 206.

12. Dick Russell, "Rush to Market," *The Amicus Journal,* Winter 1987, p. 30.

13. Doyle, *Altered Harvest,* p. 267.

14. Doyle, p. 214.

15. Russell, "Rush to Market," p. 29.

16. Doyle, *Altered Harvest,* p. 265.

17. Doyle, p. 28.

18. Doyle, p. 33.

19. Russell, "Rush to Market," p. 36.

20. Russell, p. 29.

21. Doyle, *Altered Harvest,* p. 214.

22. Bill Gordon, "Produce Buyers Prefer Organic, Group Claims," *San Francisco Chronicle,* September 22, 1988.

23. *Wall Street Journal,* January 26, 1989.

24. Nabhan, *Enduring Seeds,* p. xxii.

25. Joan D. Gussow and Katherine L. Clancey, "Dietary Guidelines for Sustainability," *Journal of Nutrition Education,* vol. 18, no. 1, 1986, p. 3.

26. *The 1988 Edition of the Seed Savers Exchange* (Decorah, Iowa: Seed Savers Exchange, 1988), p. III.

27. Michael F. Jacobson and Sarah Fritschner, *The Fast-Food Guide* (New York: Workman Publishing, 1991), p. 16.

28. Based on data provided by the National Restaurant Association from its *1989–90 Food Service Industry Pocket Fact Book.*

29. Based on data from *1989–90 Food Service Industry Pocket Fact Book.*

30. Doyle, *Altered Harvest,* p. 191.

About the Author

Martin Teitel has traveled around the world a dozen times in the past twenty years in pursuit of his environmental and human-rights work. Since 1981 he has headed the C. S. Fund. He also directs the Fund's Minor Breeds Conservancy and the Center for Seven Generations, which conducts research and education concerning heirloom seeds and other environmental issues.

Martin Teitel has a Ph.D. in philosophy from the Graduate School of the Union Institute. He is married to Mary J. Harrington and has three children.

About the C. S. Fund

The C. S. Fund is a national foundation that makes grants and operates programs to promote survival. Since its inception in 1981, the C. S. Fund has made more than $16 million in grants to preserve biological diversity, reduce toxic substances at their source, protect the ability of people to speak and act in diverse ways, and further international security.

The C. S. Fund also operates a conservancy for the preservation of rare and endangered species of farm animals and a program of biodynamic intensive gardening at the Fund's Center for Seven Generations. The C. S. Fund also maintains programs for garden apprentices and conservancy interns as well as a conference program.

The royalties from this book will help support the C. S. Fund's activities. C. S. Fund, 469 Bohemian Highway, Freestone, CA 95472.